SYMPOSIUM OF
NORTH EASTERN
ACCELERATOR PERSONNEL

SYMPOSIUM OF NORTH EASTERN ACCELERATOR PERSONNEL

Florida State University
September 28 — October 1, 1987

Editor
K R Chapman

World Scientific
Singapore • New Jersey • Hong Kong

Published by

World Scientific Publishing Co. Pte. Ltd.
P.O. Box 128, Farrer Road, Singapore 9128

U. S. A. office: World Scientific Publishing Co., Inc.
687 Hartwell Street, Teaneck NJ 07666, USA

**SYMPOSIUM OF NORTH EASTERN
ACCELERATOR PERSONNEL**

Copyright © 1988 by World Scientific Publishing Co Pte Ltd.

ISBN 9971-50-525-8

Printed in Singapore by General Printing Services Pte. Ltd.

FOREWORD

S.N.E.A.P., the Symposium of North Eastern Accelerator Personnel, perhaps at first glance a strange name for a meeting held in the South Eastern corner of the country and attended by delegates from all over the U.S.A. and from many overseas laboratories. The name is retained out of respect for the tradition and for the Canadian group who initiated the series and nurtured it for the first four years and five meetings.

The series started as a small informal meeting of the accelerator personnel from Chalk River, McMaster and Montreal and has grown to an international attendance of approximately 100 people. The sixth meeting in 1972 was the first held outside Canada when it was hosted at Florida State University. Since then it has been held at Universities and National Labs located throughout the country. A number of companies have exhibited their products and services at SNEAP in recent years adding considerably to the information available.

We are most grateful to the support of the corporate sponsors who provided refreshments during the symposium breaks and at the social functions. The conference center personnel and especially Carol Lockridge worked hard to make the meeting a success and have our sincere thanks. The trip to "Birdsong" and "Wakulla" was a great success and the entertainment by Dale and Linda Crider was enjoyed by all. Traditionally SNEAP has had a ladies program and delegates are encouraged to bring their wives. It was organized this year by Sally Chapman and the program was well attended and received many favourable comments.

Since its inception the proceedings of SNEAP have been recorded and transcribed, this involves considerable work but as much of the most valuable information is presented during the discussion periods, this is vital if the full advantage is to be obtained. This year, for the second time, we used the services of a court reporter and backed this record up by recording. This has worked quite well. The chairmen of several of the sessions edited the proceeding they chaired and this assistance was invaluable. Unfortunately the transcripts from the court reporter of many sessions were seriously delayed and these have been edited at F.S.U. in the interest of expediency.

I should like to express my sincere thanks to the many people who contributed to make this meeting a success and especially Carol Montague who organized the refreshments on the evening of the lab. tour and Lois Crew who completed the necessary typing in a very efficient and accurate manner.

K.R. Chapman

Tallahassee
February 1988

CORPORATE SPONSORS

Allied Signal Inc.
Dowlish Developments Ltd.
E.N.I. Inc.
Glassman High Voltage Inc.
G.M.W. Associates/Danfysik
Kali-Chemie Corporation
Kinetic Systems Corporation
Koch Process Systems Inc.
National Electrostatics Corporation

Exhibitors

Dowlish Developments Ltd.
E.N.I. Inc.
General Ionex Corporation
G.M.W. Associates/Danfysik
Kali-Chemie Corporation
Kinetic Systems Corporation
Koch Process Systems Inc.
Megavolt Ltd.
National Electrostatics Corporation
Peabody Scientific
Physicon Corporation
Sealey Instrument Co. Inc./Granville Phillips

ORGANIZING COMMITTEE

K.R. CHAPMAN   E. MYERS

SNEAP

Past, Present, and Future

Year	Sponsoring Institution	Participants
1968	Universite de Montreal	4
1968	McMaster University	26
1969	Chalk River Nuclear Laboratories	27
1970	Universite de Montreal	29
1971	McMaster University	34
1972	Florida State University	42
1973	Brookhaven National Laboratory	64
1974	Queen's University	32
1975	McMaster University	48
1976	Florida State University	44
1977	Los Alamos National Laboratory	53
1978	Oak Ridge National Laboratory	80
1979	University of Pennsylvania	63
1980	University of Wisconsin	83
1981	McMaster University	52
1982	University of Washington	76
1983	University of Rochester	84
1984	S.U.N.Y. at Stony Brook	112
1985	Argonne National Laboratory	138
1986	University of Notre Dame	113
1987	Florida State University	99
1988	*Yale University*	*TBA*
1989	*Oak Ridge National Laboratory*	*TBA*

CONTENTS

Foreword   v

Welcoming Address
    D. Robson   1

SESSION I
    Chairman: R. Lingren

Status of the 25 URC Accelerator
    N.F. Ziegler, G.D. Mills, M.J. Meigs,
    R.L. McPherson, R.C. Juras, C.M. Jones,
    D.L. Haynes and G.D. Alton   5

Report on 20 MV Tandar
    H. Gonzalez   11

SESSION II
    Chairman: H.R. McK. Hyder

Present Status of the Construction of the Strasbourg
35MV Vivitron Tandem
    M. Letournel and the Vivitron Group   19

Vivitron Construction — Estimated Time Planning
    J. Heugel   40

Comments on the Vivitron Beam Optics
    D. Larson   47

SESSION III
    Chairman: B. Schmidt

Voltage Tests on the Yale ESTU with Portico
    H.R. McK. Hyder, J. McKay, P.C. MacD. Parker
    and D.A. Bromley   57

SESSION IV
    Chairman: H. Jansen

New Laboratory, New Accelerator, New Challenges
    J.R. Tesmer, C.R. Evans and M.G. Hollander   77

x

SESSION V
Chairman: E. Berners

Tutorial on Computer Control
R.C. Juras   93

SESSION VI
Chairman: J. McKay

Business Meeting (This is recorded at the end
of the proceedings)
Discussion of Laboratory Reports   133

SESSION VII
Chairman: M. Letournel

Glue-Bond Failure in Dowlish FN Tandem Beam Tubes
R. Bundy, R. Lefferts, J. Preble, C. Purzynski,
J.W. Noe and H. Uto   145

Mechanical Failure of Accelerator Tubes
H.R. McK. Hyder and L.R. Fell   157

Mechanical Tests of Accelerator Tubes
H.R. McK. Hyder   161

SESSION VIII
Chairman: C. Adams

Optical Pumping in the Heidelberg Polarized
Heavy Ion Source (PSI)
H. Jaensch   179

Ion Source Work at Simon Fraser University
I. Iyer   189

SESSION IX
Chairman: L. Rowton

Report on Sao Paulo Accelerator
E. Pessoa   203

Rejuvenation of Monocast Nylon Laddertron Links
R. Bundy, R. Lefferts, J.W. Noe, J. Preble
and A. Sullivan   211

SESSION X
Chairman: K. Chapman

Developments Concerning the NEC High Gradient
Accelerating Tube Assemblies
G.A. Norton   235

Belt Charge Compensation for a Van De Graaff Accelerator
H.C. Evans, H. Jansen and G.C. Bate   246

SESSION XI
Chairman: D. Storm

Status of the Atlas Positive-Ion Injector Project
R.C. Pardo, R. Benaroya, P.J. Billquist,
L.M. Bollinger, B.E. Clifft, P.K. Den Hartog,
P.J. Markovich, J.N. Nixon, K.W. Shepard
and G.P. Zinkann   261

The Florida State University Superconducting Linac
E.G. Myers, J.D. Fox, A.D. Frawley, P. Allen,
J. Faragasso, D. Smith and L. Wright   281

The Current Status of the Kansas State University
Superconducting Linac Upgrade Project
T.J. Gray   292

SESSION XII
Chairman: T. Gray

New Fast Tuning System for the Atlas Accelerator
G. Zinkann   309

SESSION XIII
Chairman: R. Pardo

Status Report on the University of Washington
Superconducting Booster Accelerator Project
D.W. Storm   321

Superconducting Heavy Ion Booster Proposed for the
JAERI Tandem
S. Takeuchi   333

SESSION XIV
Chairman: J. Preble

University of Washington Linac Cryogenics:
Control, Performance, Maintenance and Safety
D.I. Will, D.T. Schaafsma and J.A. Wootress   345

Business Section   369

Laboratory Reports   383
    University of Aarhus
    Brookhaven National Laboratory Tandem Facility
    Kellogg Radiation Laboratory
    Chalk River Nuclear Laboratories
    Florida State University
    University of Iowa
    Kansas State University
    Lawrence Livermore National Laboratory
    Los Alamos National Laboratory
    McMaster University
    University of Notre Dame
    Ohio University
    Oak Ridge National Laboratory
    Queen's University
    University of Rochester
    Stanford University
    State University of New York, Stony Brook
    Triangle Universities Nuclear Laboratory
    Hahn-Meitner-Institute
    University of Washington Nuclear Physics Laboratory
    Whiteshell Nuclear Research Establishment
    Yale University

Appendix 1: Attainable Ion Beams — Negative Sources   445

Appendix 2: Suppliers of Special Items   451

Current Member Laboratories   460

SNEAP 1987 Participants   464

SYMPOSIUM OF NORTH EASTERN ACCELERATOR PERSONNEL

WELCOMING REMARKS

Dr. Don Robson

Welcome to the Sunshine State. It's my pleasure to make a few remarks to open this 21st SNEAP meeting. I didn't know there was going to be a court reporter here. I was going to tell a few sleazy jokes, but I'll have to forget most of those. That will at least help us get back on time.

For the past 20 years I never knew what SNEAP stood for. Everybody called it a SNEAP meeting. I decided it stood for the Society of Nuclear Engineers and Physicists, which is not true. Luckily Ken Chapman provided me with the history of the organization and pointed out the actual composition of SNEAP, so my initial remark is: Welcome to the northeast part of the Florida panhandle.

Actually some years ago some of us sat around and talked about having a symposium on low energy accelerators in the Southeast. That leads unfortunately to the acronym SLEASE.

Since its inception in 1968 this meeting has grown from about 26 participants to over 100. I think there are about 100 here, but they didn't all make it by nine o'clock, and involves participants from all over the world.

Florida State University was the first U.S. university to host this meeting in the year 1972. Fortunately for all of us involved in accelerator operations the three original participants in Canada not only had a good idea in starting SNEAP, but they also hosted the first few meetings.

This is the third time Florida State University has hosted a SNEAP meeting. I am glad to see a number of older Florida State people here who I haven't seen in some cases for 13 years.

I think this is a very timely time to have this meeting at Florida State. The tandem group have just completed building the FSU superconducting afterburner. It's not only operating, but it was built, dedicated, and came in on budget. You will all have a chance Wednesday evening to visit these facilities.

One of the hallmarks of this meeting so far as I can tell from discussions and reading the history of it is that there is a large amount of informality. Those of you wearing ties can take them off at the appropriate time, as you won't really need them at a SNEAP meeting, certainly not in Tallahassee.

I had this informality somewhat verified last week. I went to a seminar, and after the seminar one of my colleagues said he was looking forward to hearing a speaker speak at this meeting. The speaker looked rather blank, and said, "Really, are you sure? I didn't know that." This informality allows any one of you to stand up and say, "I have a problem. Does anyone have a solution?"

Perhaps some of our other physics meetings, particularly in theory where you find it cut-throat these days to go to meetings, could learn from the SNEAP format and remove one-upmanship from the physics meetings in favor of a genuine learning process.

Some of you are old-timers. You even have gray hair. Of course I don't, but there are ways around that. The experience of the old-timers is an invaluable resource and needs to be passed on, even if the solution turns out to be simply saying, "Sorry, you will have to learn to live with that," and I notice that's often the case, not just on accelerators but in a lot of physics research itself.

As a group the builders and operators of accelerators are the real foundation on which the field of nuclear physics stands today. With this in mind I would like to urge theorists in particular to think twice before they yell to their colleague, "Where is the data?" Most of you have already become immune to your professorial colleagues running into the lab shouting the nuclear analogue of the Wendy ad, "Where's the beam," which means no matter how often the users of accelerators yell at you it is clear they know your vital role. It's meetings like this which will ensure a continuation of the sound foundations which are being built by all of you.

Finally, let me warmly welcome you again to Tallahassee for the 21st SNEAP meeting. Judging by the program I see we are in for a very stimulating week in science and in entertainment. Thank you.

SESSION I

STATUS OF THE 25URC ACCELERATOR

N.F. ZIEGLER, G.D. MILLS, M.J. MEIGS, R.L. McPHERSON,
R.C. JURAS, C.M. JONES, D.L. HAYNES, G.D. ALTON

Oak Ridge National Laboratory,* Oak Ridge, Tennessee 37831

Reliability of the 25URC accelerator continued to improve in FY87 with only 253 hours of unscheduled maintenance. Operation over the past five years is summarized in Table I. A new record of almost 4600 research hours was achieved in FY87 even with a scheduled outage of 2320 hours. Most of the scheduled outage resulted from the installation of new accelerating tubes of the compressed geometry configuration(1). Four unscheduled tank openings were required during the year, but only two of these interrupted the experimental schedule. Tank openings were required by the following failures: A vacuum leak in an ion pump valve bellows; a failure in the terminal charging system; a power supply for an electrostatic quadrupole failed; a flexible coupling in a rotating shaft failed. The two rotating shafts in the accelerator have accumulated about 44,000 hours of operation, and the charging chains have operated about 38,000 hours. A summary of shaft bearing life is contained in Table II.

Replacement of the accelerator tubes has proceeded in stages. First, compressed geometry tubes were installed in two units of the accelerator and tested. Then new tubes were installed in the top third of the accelerator in November, 1986. In mid-September,1987, installation of the remaining two-thirds commenced. It is anticipated that installation and initial testing will continue until January, 1988.

In conjunction with the tube replacement, new corona points and holders are being installed in the accelerator. A new holder has been designed which will allow replacement of point assemblies without removal of the holders from the tubes. The original holders were attached to tube electrodes with a screw which would occasionally loosen allowing the point assembly to "droop" onto the lower point thus shorting the tube gap. The new holder will eliminate this problem. A drawing of the holder is shown in Figure 1. A new adjustment procedure will be employed to set the corona point gaps using a constant-current high-voltage power supply . The gaps will be adjusted to produce equal voltages (+-10%) in air with a selected corona current.

*Operated by Martin Marietta Energy Systems, Inc. under contract DE-AC05-840R21400 with the U.S. Department of Energy.

Seven new ion species were provided to experimenters during FY87. These were Ca48, Cu65, Ge76, Se78, Se80, Se82 and Sn112. The ellipsoidal geometry ion source(2) has been used since January, 1987, and has proven reliable and easy to operate. This source has proven especially advantageous for providing beams of rare isotopes since the probe samples can be very small.

The beam chopper installed in FY86 was used successfully in FY87. A replica of the beam pulse obtained from a multi-channel analyzer is shown in Figure 2. Initially, adjustment of the phases and amplitudes for the chopper and buncher was quite tedious. However, a capacitive pick-up installed in the rotating beam line eliminated the problem- at least for average beam currents greater than about 10 nanoamperes. A secondary emission detector with a microchannel plate electron multiplier has been constructed and will be installed shortly. The detector is similar to a Stony Brook design and should provide tuning criteria for beams smaller than 10 nanoamperes.

A prototype recirculating gas stripper has been constructed and partially tested. This device uses a turbomolecular pump to circulate the stripper gas. At present, it appears that some additional baffling of the stripper tube will be necessary to reduce gas flow to contiguous beam lines.

An air-SF6 separator is being developed. Since the air in a liquid storage system is found primarily in the SF6 vapor phase, air can be concentrated in one of our three storage tanks by judicious transfers between the three tanks and the accelerator vessel. The vapor in the air-rich tank then serves as feed gas for the separator. The air-SF6 ratio is further enhanced in the separator by cooling the mixture to condense SF6. The remaining air-rich vapor escapes from the separator at constant pressure while the liquid drains from the bottom.

A plan for improving the 25URC control system has been initiated. It includes additional control knobs (five versus the present three), more assignable meters, relocation of oscilloscopes in the console, and BPM modifications. A multi-channel BPM display will be constructed along with additional BPM's and remote gain control of the preamplifiers.

(1)   Jones, C.M., et al, Proc. 7th Tandem Conf., Berlin, April, 1987, to be published in NIM
(2)   Alton, G. D., Proc. Eleventh Symposium on Ion Sources & Ion Associated Technology, Tokyo, Japan (1987) 157

TABLE I
SUMMARY OF 25URC OPERATION

FISCAL YEAR	1983	1984	1985	1986	1987
RESEARCH HOURS	2969	3172	3904	4162	4591
SCHED. OUTAGE	2789	2120	2073	2396	2320
UNSCHED. MAINT.	776	1677	874	385	253

TABLE II
ROTATING SHAFT BEARING LIFE
1980 - 1987

OPER. INTERVAL (1000 hrs.)	0-5	5-10	10-15	15-20	20-25	25-30	30-35	35-40	40-45
NO. BEARINGS REPLACED	14	7	6	8	8	1	8	7	3

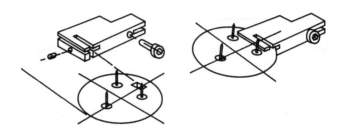

Fig. 1   New Corona Point Holder

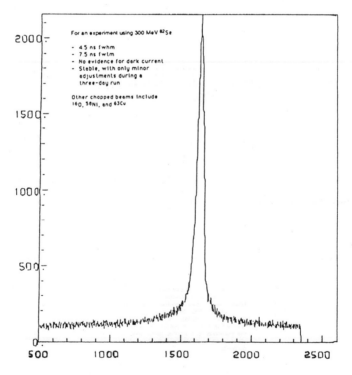

Fig. 2   Replica of Chopped Beam Pulse

Discussion Following N. Ziegler Paper

LINDREN: Are there any questions for Norval?

HYDER: Have you in fact made measurements with a sorption trap in the backing line of the recirculating stripper system?

At Oxford five years ago when I made bench tests of a system like this I decided to try putting a sorb in the backing line to trap any hydrocarbons from the turbo molecular pump. Because one has a reduction in the gas loss of a factor of about 50 over a conventional stripper, the time constants of the system are 50 times longer than for a conventional stripper, and they depend on the volume of the backing line.

If that volume is large, you could end up with a device which is almost uncontrollable, because once gas is in that volume it circulates around and around for minutes.

Furthermore, if the sorb has a pressure which is at all dependent on temperature, then one can get gas release from the sorb which can't be controlled. That again makes the control of the target thickness more difficult.

Eventually I was forced to conclude one would have to do without anything in the backing line which increased the volume or could release gas.

In fact, I also found when running a new turbo molecular pump the gases desorbed from the lubricating grease were enough to provide more than the required total flow of gases for perhaps 10 to 100 hours. We hope to put a similar system in at Yale shortly.

ZIEGLER: If you are confident enough to put it in at Yale, we can probably proceed.

LARSON: If you bypassed from the turbo pump outlet back into the inlet with a valve of some kind, which could probably be of a very simple nature, a butterfly, it's conceivable one could control the gas pressure in the stripper tube in an independent way without using feed gas, and this could perhaps even be automated, so that it would regulate the stripper gas pressure.

HYDER: I think it will still need gas eventually, you can't pull yourself up by your own boot straps.

LARSON: I didn't mean there was no need for additional feed, but one would not control the pressure using the feed gas.

HYDER: I think that's a useful suggestion. Initially the

thought was that one would need all the speed of the turbo and more, in order to get a significant improvement. It turns out that the improvement is very substantial, and therefore one can possibly afford to lose some in order to get better regulation, which is what you are saying?

LARSON: Yes.

JANSEN: You talked about the BPM display using a sampling system. Why not use four very small oscilloscopes independently, which is what we are doing?

ZIEGLER: That's certainly one option. It's difficult to look at four scopes at the same time.

JANSEN: Four may be a little difficult.

ROWTON: What is the minimum beam current you can see in your capacity pickup system?

ZIEGLER: I think it's around 10 nanoamps, average; is that right, Martha?

MEIGS: Maybe more, especially the newer ones.

LINDREN: Thank you, Norval.

REPORT ON 20 MV TANDAR - ARGENTINA - H. GONZALEZ

(Transcribed from the tape of this session)

Our machine is a 20 U.D. Heavy Ion Accelerator. In the last SNEAP at Notre Dame, we reported that we had started with nuclear physics experiments. In those days, we planned to stop the accelerator testing and make use of the machine in the way that it was at that time. We had planned to increase the voltage in the machine during this year. Unfortunately or fortunately, we do not know, we have been using the machine for research during the whole year in 1987 so no significant changes were made to our accelerator during this period. It means that we are still running with voltages up to seventeen (17) megavolts, with operation of fourteen (14) or fifteen (15) megavolts for experiments. We have been running beams of different kinds of ions such as protons, lithium, carbon 12/13, oxygen 16/18, florine, sulfur, nickle, iodine, gold and we have had the beam on target for about forty (40) percent of the available time for experiments during this year.

The main problems we have found during this period of operation were failures in the vacuum system. Twice we had failure of the needle valve in the gas stripper with

big quantities of nitrogen flowing into the tubes. This
is something that until now we do not know exactly what
happened with the valve so we have decided not the use
the gas stripper. Anyway, we were mostly using the carbon
foils until we could determine exactly what happened with
this part. In 1986, we had some problems with the Camac
system which we have solved and it is running okay now.
We also found some very bad shaft couplings in many units
in the rotating shaft. We had to repair them and now we
have new ones and we are waiting for a tank opening to
replace them. We also have had problems with the drive
belts for the charging drive motor. We had changed them
for polyurethane belts with steel cores which were not
recommended for use in that machine so we went back to
the original ones, the ones of neoprene and they run very
well.

The plans we are now having are the same that we had in
the last SNEAP. We are trying to increase the voltage in
the machine by maybe next summer or March probably. We
plan to stop the machine for experiments and we will
condition the machine and try to reach, in the first
step, seventeen (17) megavolts for stable operation and
then we will try to reach twenty (20) megavolts. We are
maybe thinking about new compressed geometry tubes, this
is something we will have to think about.

Discussion Following H. Gonzalez Paper

LINDREN: Does anyone have questions?

ZINKANN: Could you expand on the problems you had with the Camac system?

GONZALEZ: Yes. We had a failure with a microprocessor in the CRT monitor in the console. We had to wait about six months to get a replacement. That was the problem.

HYDER: Could you elaborate on the problems with the metal core drive belts? What exactly was the problem?

GONZALEZ: The problem was that this kind of belt started developing some cracks. Maybe the problem originated in the way they were made. They are made from very wide rows, and when you cut them to the size you need, I think you induce some small damage in the fibers in the belt, then it develops cracks across the belt. That was the problem we had.

I think also that it is not recommended one use this kind of belt, because the steel cords could break into small bits, and that is one of the last things you want in the machine. So that's why we went back to neoprene ones. Is that clear?

HYDER: It's clear, but I am still puzzled. Unless the belts are very badly worn, the metal is entirely within the plastic sheathing, so I don't see that one is going to get material, foreign bodies, in the tank other than the normal neoprene unless the belt wear is extreme.

GONZALEZ: Well, when the belt gets broken, that is an example of extreme wear.

LINDREN: How many broke up on you?

GONZALEZ: We broke four of them.

LINDREN: I'd like to know why you went to steel cord. What was the trouble?

GONZALEZ: Because we thought they would last longer. It was not that way.

LINDREN: I think perhaps everyone who has a Pelletron has belt dust throughout the machines, and that's the reason you want to find something else.

I have a couple of questions I would like to know about the

Argentine machine. Are you using corona points, or are you planning on going to resistors?

GONZALEZ: We are still using open corona points, the original design from NEC.

LINDREN: The other question has to do with conductive sheaves. Have you tried those?

GONZALEZ: Not yet. We are planning to do it.

MC KAY: I notice you mentioned bearing failures on the power shafts.

GONZALEZ: Not bearings. Couplings. Flexible couplings.

MC KAY: Norval mentioned they had problems at ORNL. I have one power shaft sitting on the floor ready to go in. I would like to hear more about the original problems, and if anyone has a comment or an observation to be made on the power shafts. We have one in the low energy end. When we ran it in air it appeared we were getting 140 volts, unloaded, and that seems rather high. I would like any comments on any aspects of that particular device. It sounds as though it could be the topic of future sessions, unfortunately.

LINDREN: At Brookhaven we have three drive shafts total, NEC design. We have not had any trouble with the shaft couplings unless we made it by someone leaving screws loose and did not locktite them in.

GONZALEZ: We have not yet decided what the problem is with the couplings, and have been examining the broken ones. It's not clear what is the cause of failure, so we have to work on that.

LINDREN: May I ask what the failure mode was? Did a rubber coupling break, or did it just loosen up?

GONZALEZ: No. It was almost broken, but it seems that in some parts of the coupling we have been having burning points.

LINDREN: Burning?

GONZALEZ: Burning points.

CROSS: In answer to John McKay's question, the output is about 120, 124 volts, in air. Our one failure. We did have one bearing failure. That may have been our own error in installation. It was only one failure. The four we reported on were spark damage to the windings.

HAYNES: In response to John, the problem at ORNL is the bolts were too long and did not bottom out. Then they backed out. It was a problem we had with a coupling Norval mentioned.

STIER: On the drive shafts, you will occasionally find a bearing that will die sooner than the others. At NEC, we assemble and run the drive shafts before shipping. When we find a bearing that overheats, we will change it, and continue the run. We can't run them so they wear out, so we do our best to check them initially.

ZIEGLER: Our shafts have about 40,000 hours on them. Some of our bearings have never been changed. One bearing has been changed three times. I think in general the shafts have proven very reliable.

GONZALEZ: We didn't change any bearing in the shafts yet.

ZIEGLER: We do have a little instrument for measuring vibration in bearings, and we record this at every tank opening. We record the vibration from each bearing. I think this is rather helpful if you have very many. It gives you a history of what is happening.

I would also like to comment on the belts. We, too, tried these metal core belts accidentally in the terminal. The belt ran up against an edge of the pulley and started fraying, and we had little metal pieces all through the accelerator. It's not a good choice.

LINDREN: I would like to get something clear before we take any more questions. What is the diameter of the vertical machine drive shafts, six inches?

ZIEGLER: Five, I believe.

LINDREN: Okay. The horizontal MP conversion uses two-inch diameter shafts. So the bearing problems are quite different.

CROSS: (directed to the NEC people) Do you align the drive shafts using a scope, and what type grease do you use in the bearings?

STIER: We do align the drive shafts with telescopes. We specified Chevron SRI type of grease. It's always difficult to get the manufacturing people to realize the significance of the lubricant. You have to be very careful. They want to give you anything.

CROSS: We have found Crytox 240-AC is a very good grease grade. When we have a belt with rollers, six rollers, we had multiple failures until we found Crytox grease, which is quite expensive--$250 for a pound of grease.

LINDREN: The grease NEC supplied us that has been working very

16

well at BNL is Chevron SRI grease. We use it in all bearings in the tank gas. We did at one time go to Crytox, hoping it was good, and it was.

CHAPMAN: What sorts of sources are you using at this time?

GONZALEZ: We have three different sources. A duoplasmatron, a Cesium sputter cone, and Alphatron. We are now using Cs sputter source for almost every experiment.

CHAPMAN: Is that the sputter source developed by Gerald Altan at Oak Ridge?

GONZALEZ: No. It's the SNIC source, supplied by NEC.

LINDREN: Any more questions? Then that will be all. Thank you.

SESSION II

PRESENT STATUS OF THE CONSTRUCTION OF THE STRASBOURG

35 MV VIVITRON TANDEM

Michel LETOURNEL and the Vivitron Group

Centre de Recherches Nucleaires
BP 20   67037 Strasbourg-Cedex (France)

INTRODUCTION

Some months ago, in Berlin, I gave a paper on the status of the Vivitron. There were not too many people attending this SNEAP. So my talk here will be quite similar to the Berlin's one, in order to inform you of what is going on at Strasbourg. The construction of the 35 MV Vivitron is in progress. At the end of 1986, the 50 m long tank was already in situ in its new building located parallel to the existing MP building. Building and tank are now finished and the construction of the generator inside the tank has already begun. Jean Heugel, who attends this SNEAP meeting, is in charge of the mounting and he will use part of my time to give you his estimated time planning for the construction. Actually, the major part of the generator components have been ordered and will be delivered soon, i.e., post insulators, column insulators, discrete electrodes and column electrodes as well as the total gas handling system. Tests have been carried out, specially in 1986, on a complete real Vivitron section, on various post insulators and on the belt charging system in the CN machine. In addition, the subject of transient in the Vivitron in the event of spark with regard to the problem of the stored energy, its dissipation and the over-voltages, have been thoroughly investigated and discussed. I remind you on the figure I the Strasbourg project to install a 35 MV Vivitron tandem in parallel to the MP. Today the 50 m long Vivitron tank (fig. 2) is lying in its vault. 1987 represents the third year of funding and the building and the tank are now completed. All the components required to build the generator (fig. 3) are now ordered. Phase I started last September with the mounting of the interior components; it will last up to the tests of the 35 MV Vivitron generator, expected early 1989. Phase II corresponds to the accelerator itself and beam is expected in 1990.

REVIEW OF THE MAIN FEATURES

To recall briefly the characteristics of the 35 MV Vivitron tandem it is a horizontal machine built out from the tank wall using post type insulators and discrete electrodes. For the outside of the central column:

Fig. 4 shows the column electrodes, the assemblies of discrete electrodes alternating with the post insulators.

Fig. 5 shows the corresponding field mapping.

Fig. 6 shows the general field mapping in a cross-section of the Vivitron terminal area.

For the interior of the central column, fig. 3 shows the column insulators, the belt running from the LE side up to the HE side, and the accelerating tube sections.

Roughly speaking this machine is the double of the Strasbourg MP 10 accelerator and is capable of withstanding 35 MV longitudinally distributed by resistor chains on 8 column sections on each side of the terminal electrode.

EXPERIMENT CN 2 ON A REAL VIVITRON SECTION

In 1983 a preliminary experiment was carried out on the Strasbourg CN machine to demonstrate the compatibility of long insulators of different composite materials in a non-subdivided column. Early in 1986, the experiment included a prototype of the final insulating column structure plates and also the final column electrodes (fig. 7). It represents a real Vivitron section. The test was carried out with a modified terminal electrode.

The column plates are cut out of plain plates of composite material made from short glass fibers and epoxy (fig. 8). The 6 column electrodes of identical conical shape are supported by the column structure and connected to an internal resistor chain. Each gap is intended to withstand 750 KV and to act as a spark gap (fig. 9). Actually, the prototype Vivitron column section easily exceed its 4.5 MV design rating, it reached 6 MV and long duration tests were achieved. The special requirement was to dry thoroughly the insulators.

TESTS ON THE POST INSULATORS

The final post insulators (fig. 10) designed by C.M. Cooke, have been tested using the CN machine. A special arrangement around the CN terminal electrode (fig. 11) reproduces electrical conditions similar to those between two discrete electrodes. Two circular electrodes establish a radial gap of 40 cm and locations are provided to allow tests of up to 6 posts at a time.

Tests have been performed easily at 4.5 MV with and without sparks. Long duration tests have been accomplished for one month at 4.5 MV and then at 5 MV for one week. Actually the posts are easily capable of holding 5 MV or more. A good illustration of breakdown at a post is shown on fig. 12, the spark at around 6 MV passed between the hoops below the CN terminal and the opposite electrode without reaching the surface of the post. All 244 posts to be installed in the Vivitron will be electrically tested with this arrangement in the

CN. Some of them will be also tested mechanically.

J. Heugel was in charge of all the tests on the insulators in the CN machine and P. Wagner made the calculations and field mapping.

STORED ENERGY

At 35 MV, the stored energy is about 400 kJ. It is mainly distributed outside the central column. The Vivitron design is essentially to allow this energy to be dissipated outside the column in the event of spark and to prevent damage to the interior of this column. Many discussions on this subject and on transients have been held. Calculations and model analyses have been done by C.M. Cooke and specially A. Staniforth. The preliminary results confirm the good behavior of this structure for the transcients due to the distributed energy and its ability to protect the different components and particularly the post insulators. More results will be available in a short time and these will be the subject of a report.

ACCELERATING TUBE

All the optics and the arrangement of the accelerating tubes have been calculated and designed by J.D. Larson (fig. 13). Overall, the ion optics are similar to an elongated MP tandem with a fixed object distance, a gridded entrance lens, an electrostatic quadrupole lens assembly in the terminal, a crossover at a second stripper in the HE column and a magnetic quadrupole doublet lens located outside the HE base of the tank. The terminal lens can be configured as a charge selector with unwanted charges stopped at (or before) the second stripper location. There are a total of 18 accelerating tubes, 9 on each side of the terminal. Fourteen tubes are 100 inches long. The tubes nearest the terminal and the HE exit are half length in order to displace tubes so that tube joints do not coincide with locations where discrete electrodes join to the column. Most tubes have 4 sections of 14° inclined electrodes in an alternating sequence. A few of the full length tubes have 6 inclined sections; these will be used where the energy gain is large. The incline pattern is designed to return the beam to the axis at the end of each tube. Tube 1 is fully inclined. The incline at the entrance is 7° and the gridded entrance lens is also tilted by 7° to conform. Half value resistors are used through the initial 7° incline section and the following 14° section. Corrective pre-steering must be applied to the injected beam to overcome the initial inclined field perturbation. the design injection energy is 250 keV but energies as low as 100 keV can be accommodated.

MECHANICAL CONSTRUCTION

Every second dead section the column is supported by 3 series of post insulators hanging from the tank roof and 2 series of post

insulators coming up from the tank bottom (fig. 14). Fig. 15 shows
the arrangement of column, posts and discrete electrodes around the
terminal. G. Gaudiot is in charge of the mechanical construction.

PRESENT STATUS

   The 50 m long tank was brought on the site in parts (fig. 16)
and assembled in the new building. There are 3 main rooms, one for
the injector, one for the accelerator itself and one for a new 90°
analyzing magnet. The tank has a volume of 1200 m³. It has been
pressure tested using water in last April at a pressure of 24 bars.
Then it has been grinded polished and white pointed. Since last
September a scaffolding has been installed inside the tank under Jean
Heugel's direction. Jean will talk now about the mounting of
the interior components and his estimated planning.

Figure Captions

Fig. 1   General layout of the Strasbourg Nuclear Center with the 35
         MV Vivitron tandem.

Fig. 2   The 50 m Long tank installed in 1986.

Fig. 3   The 35 MV Vivitron components.

Fig. 4   The column electrodes and discrete electrodes.

Fig. 5   Field mapping, across an insulator.

Fig. 6   Field mapping across - section of the Vivitron terminal area.

Fig. 7   The CN experiment.

Fig. 8   Composite panels assembly for the CN experiment.

Fig. 9   Field mapping along the column section for a test in the CN.

Fig. 10   A 40 cm long post-insulator.   Management for mechanical
         test.

Fig. 11   Special management around the CN terminal electrode for
         post-insulators tests.

Fig. 12   Breakdown at a post-insulator at 6 MV.

Fig. 13   Vivitron accelerating tube.

Fig. 14   Components arrangement: longitudinal and cross section.

Fig. 15   CAO picture of the Vivitron components assembly.

Fig. 16   The tank was transported in several parts and assembled in
         Strasbourg.

24

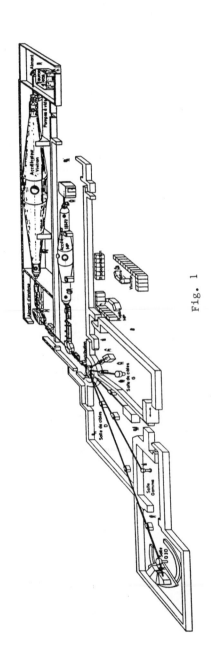

Fig. 1

Fig. 2

26

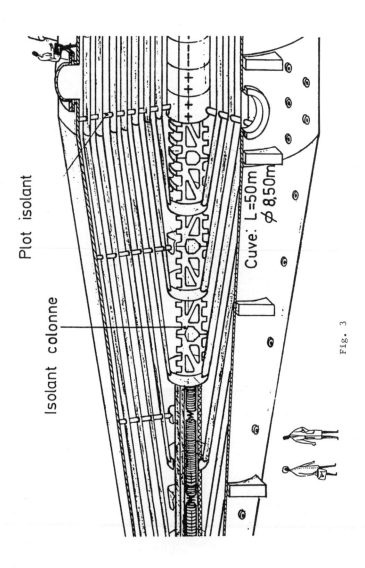

Plot isolant

Isolant colonne

Cuve: L=50m ⌀8,50m

Fig. 3

27

27

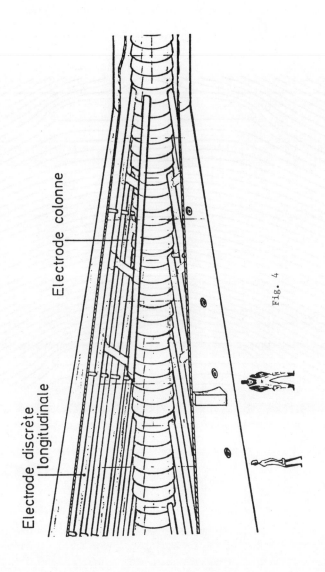

Electrode discrète longitudinale

Electrode colonne

Fig. 4

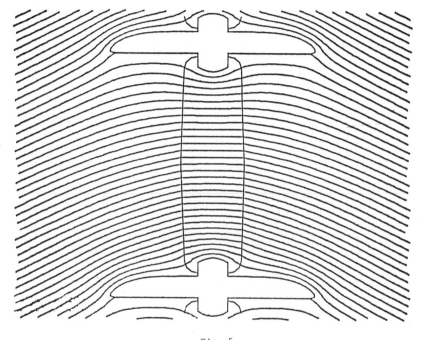

Fig. 5

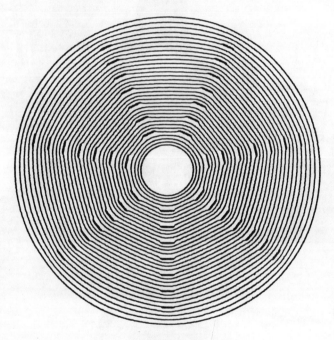

Fig. 6

Fig. 7

Fig. 8

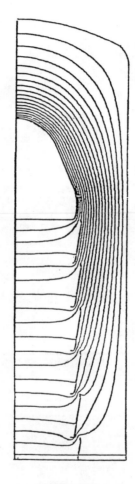

Fig. 9

Fig.10

Fig.11

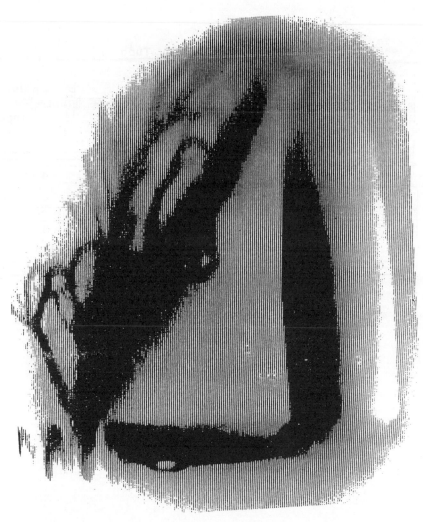

Fig. 12

VIVITRON 35MV-Accelerating Tube

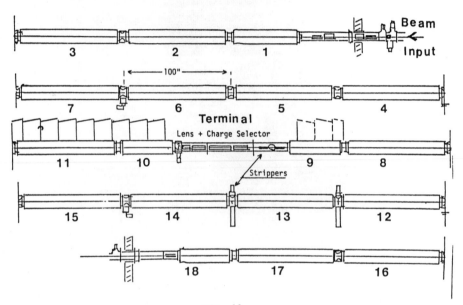

Fig. 13

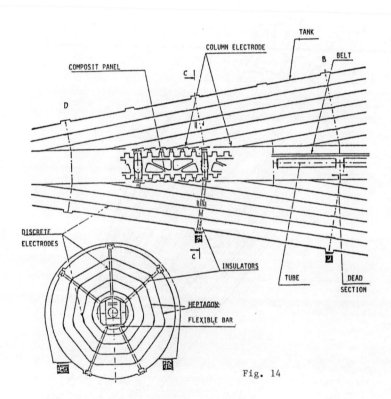

Fig. 14

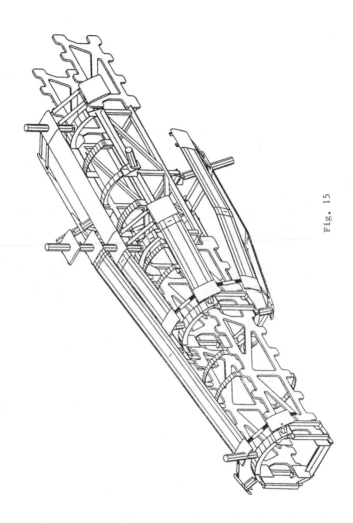

Fig. 15

Fig.16

VIVITRON CONSTRUCTION

ESTIMATED TIME PLANNING

J. HEUGEL

INTRODUCTION

PLANNING FOR THE VIVITRON CONSTRUCTION:

- estimated time planning

- made by Macintosh "MacProject"

REMARKS:

- Vivitron = electrostatic prototype machine
  issued from Michel's (crasy) ideas

- Strasbourg group ---> hard working to arrange an
  electrostatic accelerator

- no experience for so a large structure assembly

====> obviously, we have to adjust the estimated time after
the first sections assembly

- beginning column structure assembly: NOV 9, 1987
(first step)

VIVITRON CONSTRUCTION TIME PLANNING

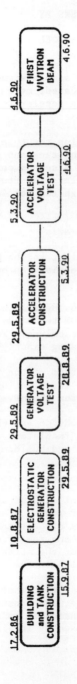

ELECTROSTATIC GENERATOR CONSTRUCTION PLANNING

DISCUSSION FOLLOWING PAPERS BY

MICHEL LETOURNEL AND J. HEUGEL

McKAY: You assure us that this is a simple structure but, having seen something one seventh as complicated, my question is: "How are you going to maneuver 100 inch long accelerator tubes into place after the voltage tests?"

HEUGEL: In the first place we will bring the tube sections through the large openings at the ends of the tank and move the tubes through the column sections into position. A second possibility is to introduce the tube sections through the large aperture at the bottom of the tank and then lift them through the portico sections onto the central axis.

McKAY: Do you move the tube along the axis?

LETOURNEL: Yes, we have a cradle, similar to the MP's. We have designed the column so that a single tube section can be removed, complete with resistors.

HEUGEL: We have had experience of removing single tubes and replacing them in the MP, but in the VIVITRON the situation is more difficult because there is less space in the column. We will wait until the whole column is built before deciding exactly what to do.

LETOURNEL: The big aperture at the bottom of the tank will be used only for single tube section changes. Installation of the complete set will be from the tank base as for the MP.

WEIL: During the assembly procedure, why do you choose to assemble from the top downwards so that the insulators are in tension, rather than starting at the bottom and building up with the posts in compression?

HEUGEL: This follows advice from Chat. Cooke that by building from the top down, we would reduce the amount of dust and foreign bodies attaching themselves to the insulator posts. The mechanical considerations favored the use of seven rows of post insulators instead of eight as in the MP, resulting in one row of posts in the vertical plane, but the conclusive reason for the arrangement was the electrical requirement to avoid dust particles on the assemblies.

LARSON: I would like to comment further that what Michel Letournel is saying is that in operation the upper posts are designed to be in tension, and as they operate in tension it is sensible to construct from the top down and hang the assembly in pieces at the appropriate moment.

HEUGEL: We will build the portico one layer at a time. We will assemble all the insulators belonging to one layer, complete that layer and then move into the next layer. Inside each layer we start at the top and finish at the bottom.

LETOURNEL: Some of the posts have been subjected to a tensile stress of 5 tonnes but they are designed to withstand 20 tonnes in tension.

BERNERS: How is it possible to run this machine with long insulators, not subdivided, in the portico assembly and even in the column without getting surface flash-over?

LETOURNEL: When we first upgraded the MP from 13 to 16 MV, we found that although we could run at 16 MV, if we had a spark at 15 MV we damaged many components inside the column, resistors and even the column insulators. So we started to think about the energy, 99% of which is stored outside the column, so as to prevent it from being released inside the column. This led to the idea of discrete electrodes which we applied to the MP with result that column damage ceased and we were able to increase the voltage to 18 MV without further damage. The VIVITRON uses these ideas, extended to the larger machine.

ROWTON: Why did you choose the Swedish belt? I though that when you tested it, it was nothing out of the ordinary.

LETOURNEL: In fact the tests have been carried out in other European laboratories and the results are OK. Mr. Arndt from Berlin may be able to comment on this. We are confident that these tests show that it is OK. We have tested one in our C.N. accelerator, satisfactorily. It is at present the only available belt which will fulfill our goal.

UNKNOWN: Will you describe briefly the VIVITRON vacuum system?

HEUGEL: We will put a large turbo-molecular pump with a speed of 1500 l/s at each end of the tank. Because the vacuum conductance of the accelerator tubes is very bad, about 1.75 l/s, we will put an ion pump into each of the unsupported dead sections. There is a problem with the space which results in ion pumps only in dead sections 2, 4, 6, 10, 12, and 14. The terminal design is not completed, but I think we will have to put cryopumps there and we also plan to install turbo pumps in the same way as at Oak Ridge, Yale and Berlin. But the details are not completely designed.

UNKNOWN: What sealing method are you going to use between the accelerator tubes?

HEUGEL: These will be standard vacuum seals. We have a double VITON "O" ring at each joint inside the tank so that we can check every joint by pressurizing with helium.

UNKNOWN: What vacuum range do you expect to reach?

HEUGEL: That is a good question. In the badly pumped tube sections we expect to have 10^{-6}Torr and better in the sections which are pumped. The problem is with the tube conductance.

ZINKANN: From a maintenance standpoint, how do you gain access to the terminal once the portico is constructed? Do you have to disassemble the structure?

HEUGEL: This is a big problem and we will put Michel Letournel inside the tank to solve it.

We have discussed several different ideas. We have to be able to move through the different layers of discrete electrodes and then, for easy access to the center terminal, we have a platform. This can be built up easily from the bottom of the tank at the center to give access to the side of the terminal. We will have to remove some of the column electrodes to get in.

ZINKANN: Would the platform be supported by the discrete electrodes?

HEUGEL: No, the platform will have its own supports. The discrete electrodes of the portico are designed to support one or two people, so that they can be used to climb up to the platform.

SOUTHON: It looks as if the electrodes join the column at a midpoint between tube dead sections. I guess that means that the tube and the column are decoupled.

LETOURNEL: That's right.

SOUTHON: Do you have separate resistor strings on each?

LETOURNEL: Yes. These chains are located on either side of the tube, not on the top or bottom as in an MP. The shape of the resistor mount is being redesigned so that it is a little different from that in the MP. The reason why the tube joints are not at the column dead sections was to answer objections raised at the start of the project design when people said that the stored energy inside the column would reach the accelerator tubes and would be too much. In order to avoid that we went to the present design, which is quite complicated. If the decision were made now, I would not use this idea, since all the subsequent test have shown that the most energy which the column must cope with due to a single breakdown is about the same as a spark in the CN. The energy is about 5KJ and neither the amplitude nor the frequency spectrum are a danger.

WILL: How does the corona regulating system work?

LETOURNEL: It will work in the conventional way.   I must confess that we have not paid too much attention to the need to modify the MP design.   One reason for this is that when you install a portico the stability of the tandem is improved.   Because of its size and the number of porticos the VIVITRON will be very stable.   Of course we need the corona loop for regulation and stability but a conventional device will be satisfactory.

WILL: Will the corona penetrate between the portico electrodes?

LETOURNEL: Yes, just as on the portico test.

GRAY: How many man years does the three year construction schedule amount to?

HEUGEL: I expect to have about fifteen people working the project.   At present there are about ten for the building work and the accelerator structure.

CHAPMAN: In April, careful measurements were being made in Strasbourg in case the tank was not truly cylindrical, because there was concern that a non-cylindrical vessel might distort as well as expending on pressurization and that this movement might exceed the yield in the support structure.   Has this problem been resolved?

HEUGEL: Yes.   The column hangs from three lines of insulators which will follow the movement of the tank.   As the bottom legs cannot follow this movement, we insert a special flexible bar between the column and the bottom insulators so as to allow for this movement.

CULVER: Will this bar also accommodate vibrations in the terminal or the column, such as the swaying that occurs in the MP when the fans are going?

LETOURNEL: Yes.   The vibration comes from the belt and the rotating shafts.   All these mechanical components are mounted at the end and we expect the effects to be less than in the MP.   The belt is only running at 10 m/s.

JANZEN: Is there only one belt?

LETOURNEL: Yes. It is 100 m. long.

STURBOIS: How long is it going to take to change a belt?

LETOURNEL: It is very easy because, as now in the MP at Strasbourg, there is nothing between the two runs of the belt. Therefore we have nothing to remove, except the motor and alternator, before sliding out the belt and replacing it. Even if the belt is continuous, like the HVEC belts, we do not have to disturb the structure.

HEUGEL: But we have to remove some column electrodes in order to remove the alternator.

COMMENTS ON THE VIVITRON BEAM OPTICS - DAN LARSON

(Transcribed from the Tape of this Session

Checked by D. Larson)

The optics of the Vivitron is extremely simple. In fact, I am
embarrassed to tell you about it. The low energy stage of the
machine is the same as an MP, really. There is a crossover about two
and one half (2 1/2) meters in front of the tube entrance. At the
tube entrance a grid is stretched across the first active electrode
in the tube. The grid prevents equi-potentials from extruding
outside the tube (or the fringing fields from coming outside the tube
at the entrance) and, therefore, cancels the natural focusing of the
tube which is exceedingly strong because of the low energy of the
injected ions and the high gradient in the tube. A lens is then
constructed at that grid by using a grounded electrode outside the
grid and some voltage applied to the grid. The focusing for the
injected beam is performed by that "gridded lens" in conjunction with
some additional focusing that occurs in the interior of the tube due
to the fact that the gradient at the entrance to the tube is not the
full gradient but is reduced to half gradient. Where gradient change
occurs a little further into the tube, there is a natural lens
action. However, the beam energy has increased by that point and
therefore the focal length of that second lens is long enough so that
it takes a combination of the lenses to produce a focus at the
stripper location in the terminal.

This is not optimum from a beam transport standpoint. This is
optimum from the point of view of maintaining voltage in the
accelerator. The optimum beam transport arrangement would be one or
possibly two lenses in the low energy stage of the accelerator such
as the Daresbury or Oak Ridge arrangement. Some smaller machines
also have lenses in the low energy column. But, those lenses take up
space and in this machine (the Vivitron) that space is being used to
reduce the gradient in the tube; that is, the tubes are made longer.
The acceptance of the machine is going to be of the order of half of
an MP acceptance so if you can get reasonable beam currents through
an MP, you should also be able to get reasonable beam currents
through this machine.

The high energy stage is also like a typical MP starting from
the stripper. Well, perhaps not like a typical MP but at least some
of the MPs. There is sufficient room in the terminal of an MP, and
many of the larger accelerators, to introduce a lens to match the
beam into the high energy stage of the accelerator. Where one is
using a second stripper, one would like to be able to focus the beam
to a crossover at that stripper location. In fact, the best match of
the beam into the high energy stage of an accelerator is to produce a
crossover somewhere in the middle. Just think of the high energy
stage as a collimator, for example. You want to focus the beam in

from the outside so it crosses over and is expanding again as it leaves the high energy stage.

There was not room, we felt, in the Vivitron terminal to have both a charge selector assembly, self contained with a focus between the stripper and some crossover point still located within the terminal, and then after that have a matching lens to accommodate the requirements of the high energy stage. Consequently, an alternate idea has evolved which is to modify the matching lens so that it becomes a charge selector. The proposed lens for this application is a rather unusual quadrupole assembly. The beam in this case is so far off center that it is no longer necessary to have four poles in the quadrupole but one can use a single pole, add mirror planes and construct the missing three poles by image charges. A schematic representation of that is indicated here (points to Fig. 1). The off-center lens elements cause dispersion to take place in the beam much like adding a dipole field in this location. Following the offset triplet, which does the charge selection, there is an additional matching lens element, a singlet, which is not a full matching lens but adjusts for some of the variations that can occur in the focusing requirements of the system. In fact, that singlet is very useful to us for reasons that I won't go into in detail. (This is in reference to alternative modes of operation which permit the focus to be adapted to different beam conditions and also allow the amount of dispersion to be changed.)

Here is another view (Fig. 2) of that portion of the machine, a little better to scale. It shows the stripper, triplet charge selector and singlet element. The 100 inch lengths (indicated in the figure) are the active portions of the tubes. The modular spacing is 111 inches.

Here are some calculated results (points to Fig. 2) for the effect of the charge selector, operated in a particular mode. (There are some alternatives as to how to operate this thing.) This calculation was done for a potential of 35 MV on the terminal. It is at a very low charge state which I think is not representative of how the machine will be operated. A central charge value of 6 has been selected for the beam of interest and this diagram shows where charges 5 and 7 are expected to go. Under these circumstances, the beam leaves the high energy end of the machine here (points to figure) and goes through the conventional high energy doublet lens to an object crossover preceding a conventional 90 degree energy analyzing magnet.

The next figure (Fig. 3) is getting a little more realistic in charge state but the calculation was done for what I consider to be, hopefully, a very unrealistic terminal voltage of 20 megavolts. Under these circumstances, you see that for a selected charge of 13, charges 12 and 14 are widely separated. Now, the process of eliminating the spurious unwanted charges, 12 and 14 in this case,

will begin immediately in the terminal with some aperture limitations
that will be in the order of one inch or so in diameter, I would
expect. Additionally, some wiping off of charge can take place at
dead sections going down the high energy stage but ultimately the
final selection point is at the location of the second stripper.
There, one can present an aperture of moderate size, something in the
order of one or two centimeters, and separate out the unwanted
charges from the selected charge.

This system has the undesirable feature that charges that are
being rejected cannot be eliminated before they leave the terminal,
at least in some cases, and will proceed down the high energy stage
until they are finally separated out at the second stripper location.

Cas examinés	Distribution Intensité-émittance	Transmission		
		BE	HE	VIVITRON
U = 20 MV, A = 130, Q = 13$^+$	linéaire	0,32	0,50	0,16
	gaussienne	0,52	0,65	0,34
U = 35 MV, A = 12, Q = 6$^+$	linéaire	0,54	1,00	0,54
	gaussienne	0,66	1,00	0,66

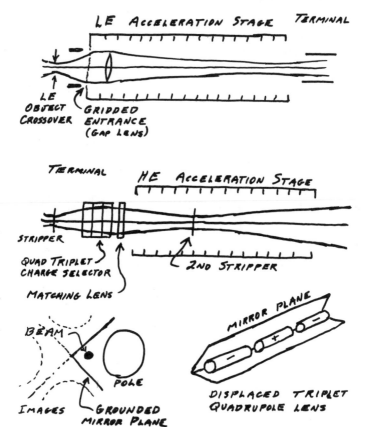

Fig. 1 Basic Indication of the Beam Optics and the Off-Set
Quadrupole.

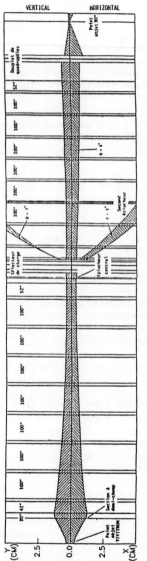

Fig. 2 Calculated Beam Envelope at 35MV Showing Dispersion of Charges 5 and 7 When Charge 6 is Selected. (R. Rebmeister CRN-VIV-34 1986)

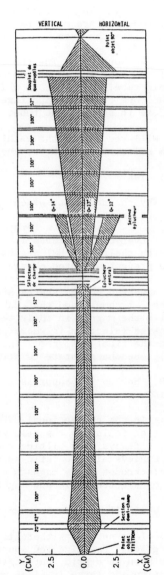

Fig. 3 Calculated Beam Envelope at 20MV Showing Dispersion of Charges 12 and 14 When Charge 13 is Selected. (R. Rebmeister CRN-VIV-34 1986)

DISCUSSION FOLLOWING D. LARSON PAPER

GRAY: We do what we can to eliminate having the beam run into the tube at any point in the tandem, but it looks like in this charge selection process you are creating a situation where you run the beam somewhere down into the tube.

LARSON: We hope not. The charges will be dispersed in the wide direction of the slot-shaped aperture of the conventional HVEC inclined field tube. So there is a lot of space in the direction of dispersion and one can place apertures at dead sections to pick off the beam before it strikes the tube.

GRAY: Will you follow the machine with a conventional charge and momentum selector to pick out the beam you want?

LARSON: Yes.

GRAY: Why the necessity of charge selection? Are you expecting that much input beam?

LARSON: Everybody believes it is a good idea. Daresbury are very happy with theirs and HMI are using one. In this case the object is to prevent unwanted charges from hitting the tube electrodes. Without charge selection this would certainly occur because the unwanted charge states are not correctly focused by the terminal lens. I should add that Oak Ridge have ideal charge selection with the magnet in their terminal. One situation that is helped by charge selection is where proliferation of charges leads to ambiguity in the final analyzer. That can certainly be helped by allowing only one charge state to strike the second stripper.

CHAPMAN: How do you suppress secondary electrons from the entrance grid?

LARSON: In the conventional HVEC inclined tube, there is a non-inclined section following the grid with external magnets to deflect the electron sideways. By tilting the entrance electrodes it is possible to achieve electrostatic suppression. At Strasbourg, even the tilted electrodes carry magnets so suppression is by a combination of electrostatic and magnetic fields.

WEIL: What lifetime do you expect for that grid?

LARSON: I do not think there is a problem with the lifetime of MP entrance grids. Positive ion sources can produce a high enough intensity to burn up a grid, but at the tandem entrance, the beam is low in intensity and large in size, not focused to a spot. The grid is inside the pressure vessel, but I have not heard of any trouble in that position.

HYDER: The grid of the Oxford Folded Tandem was removed after 30,000 hours for inspection.   There was no visible deterioration. The is also the common experience of most MP users.   At the tube entrance the beam diameter is large enough and the intensity low enough that the life is very long, in contrast to grids used immediately outside ion sources where the effect of neutral flux may be to limit life to a few hundred hours.

SOUTHON: How do you overcome the problem of unequal focusing in the x and y planes at the point of gradient change?

LARSON: The tube has two circular apertures at that point.   At least that was my conceptual design.   I do not know the details of the present tube design, but such a tube has been operating at Strasbourg for two years.

SESSION III

VOLTAGE TESTS ON THE YALE ESTU WITH PORTICO

H.R. McK. Hyder, J. McKay, P.D.MacD. Parker and D.A. Bromley

A.W. Wright Nuclear Structure Laboratory
Yale University, New Haven, CT 06511

1. INTRODUCTION

MP-1 tandem accelerator at Yale has been converted to an ESTU tandem with a design voltage of 20 MV. The MP-1 accelerator was taken out of service in May, 1985 and by January, 1987 the new machine was ready for high voltage tests without portico and without accelerator tubes. The main features of the ESTU and the results of the first series of voltage tests have been reported[1]. Since then, the portico has been installed, and a further series of voltage tests have shown that the electrostatic behaviour of the column and portico is satisfactory and consistent with the design aims. The results of these tests are reported here together with some conclusions about the behaviour of the portico and the requirements of a voltage stabilizer system.

2. ESTU DESIGN

The accelerator is described in ref. 1 and illustrated in figs. 1 and 2. The effect of the portico on the radial electrostatic fields is shown in fig. 3. By adding a fifth section to the original low-energy and high-energy columns of MP-1 and by the use of extended accelerator tubes with 88 active insulators, a terminal voltage of 20 MV or more can be expected from the ESTU, corresponding to a column gradient of $\sim$ 4 MV/section or 45 kV/inch on the accelerator tubes. In order to achieve this voltage without exceeding the conservative value of 15 MV/m for the macroscopic radial field it would be necessary to enlarge the terminal diameter to 1.4m and to allow the tank diameter to increase to 8m or more if an intershield was not employed. By introducing an intershield at 0.6 V_T, where V_T is the voltage between terminal and ground, a standard terminal shell can be used in a tank of 7.6m diameter to give a design in which a terminal voltage of 27 MV corresponds to a maximum macroscopic field of 15 MV/m. The amount of stored energy in an accelerator as large as the ESTU makes such a safety margin highly desirable.

A conventional intershield, consisting of a uniform cylindrical electrode surrounding the terminal and inner column, acts as a Faraday cage decoupling the inner and outer parts of the accelerator so that disturbances in one section do not necessarily trigger breakdowns in the other. The large area of such an intershield results in a high capacitance and large stored energy with consequent risk of damage to the column near the point of attachment. By contrast the open structure of the **portico** proposed by Letournel and operating at Strasbourg[2], reduces the stored energy and hinders high frequency current flows because of the relatively high impedance of the portico panels.

Under static conditions, the **portico** controls the voltage distribution between terminal and tank in much the same way as a solid electrode. Careful detailed design of the panels and transverse stiffening bars minimizes the area of electrode exposed to high field under static conditions. However, if the voltage on either the inner or the outer column changes rapidly, transient field distributions arise which couple the inner and outer volumes, impressing high surface fields onto parts of the electrodes which are not stressed as severely in static operation, and leading to total voltage collapse.

The portico is therefore expected:

- (i)   to increase the maximum voltage by reducing the radial fields
- (ii)   to change the behaviour of the accelerator during spark discharge or other voltage excursions
- (iii)   to interfere with the operation of the central generating voltmeter and CPU and to modify the effect of the corona probe draining current from the terminal.

3. TESTS WITH THE PORTICO

3.1 Measurement System

Temporary instrumentation was used during these tests. In addition to measuring the column current at the LE and HE bases, the currents entering the columns from the terminal and the currents at the points of attachment of the portico were estimated by the use of simple current integrators optically linked to the outside. These consisted of a 90 V stabilizer neon in parallel with a capacitor of known value. As current flows into the circuit the voltage on the capacitor rises until the neon fires, allowing the process to repeat. The light from the neon is focussed onto an adjacent port fitted with a photomultiplier, pulse-forming circuit and ratemeter.

Generator voltmeters were installed at three positions along the tank parallel to the axis facing the dead-sections between tubes 2 and 3, 3 and 4, and 7 and 8. A fourth instrument, opposite the center of the terminal, was located so as to face a gap between portico panels. The voltmeters opposite the ends of the portico were assumed to respond solely to the voltage on the portico. The response of the central voltmeter to changes of potential between portico and terminal was measured by shorting the inner column. This changed the reading by only 5% showing the need to mount a generator voltmeter in the terminal, or perhaps the portico, in order to get reasonable sensitivity.

In addition to the generator voltmeters and column current meters the Pelletron upcharge, Pelletron ripple and portico ripple were also observed. Visual observation of the column was possible through a total of fifteen ports and three Polaroid cameras were used to record spark patterns.

3.1.1 Tests without radiation sources

Each of the three runs began with tests at low gas pressure and without radiation sources. As the charging currents were increased, the ratios of generator voltmeter readings and column current readings were observed and compared with the current imbalance to detect the onset of corona or other leakage. At low voltages, good charge balances were observed, and the column gradients were sensibly uniform. Above a threshold which depended on gas pressure, leakage currents began to flow, distorting the column gradient. The presence of the portico provides a number of alternative paths for leakage current, as indicated in fig. 4. As expected, relatively high voltages were obtained at low gas pressures, but as the gas pressure was increased the rate of increase of voltage slowed down, and it became necessary to condition for lengthy periods and, eventually, to introduce radiation sources. No prolonged efforts were made to establish maximum voltages at low pressures, but it was comparatively quick to reach 7.5 MV at 2 bar. After conditioning for 30 hours a terminal voltage of $\sim$ 15.5 MV was reached at 6 bar with no radiation sources.

3.1.2 Tests with radiation sources

The ESTU has two radiation sources on the center line of the tank: a 2-Ci source near the top and a 4-Ci source near the base. For the present test the sources were screened by hemispherical aluminum covers 6 inches in diameter which protruded 3/4 inch above the surrounding surface and which contained a lead screen designed to absorb most of the radiation in the 90° cone pointing towards the tank center.

When a source is inserted leakage currents flow from the terminal to the tank, from the portico to the tank and, to a lesser extent, from the terminal to the portico and from the column near the terminal to the tank and the portico. The precise origin of these latter currents can not be ascertained using only meters in the terminal and dead-section. The magnitude of these leakage currents increases with source strength and with terminal voltage in rough proportion. Apart from any influence on maximum voltage, these currents result in an increase in the gradient of the column inside the portico, especially near to the terminal.

The variation of leakage currents with source strength and terminal voltage is shown in fig. 5. The increase in inner column gradient due to leakage currents from the portico can be detected as a change in the ratio of generator voltmeter 4 (opposite the center of the terminal) to generator voltmeters 2 or 6, which measure only the voltage on the portico. Because of the low sensitivity of the GVM in this configuration, the measured effect is small, but it is significant in terms of the uncertainties. This change in ratio has been used to confirm direct measurements of current leaving the terminal for the inner column and of the current to the portico from the inner column. Knowing the column currents at both ends of the inner and outer columns it is then possible to calculate the true terminal voltage. The maximum terminal voltages, determined in this way, obtained during operation with and without sources are listed as follows:

No sources	(leakage current 1-2 μA)	V_T = 17.5 MV
2-Ci source	(leakage current 8 μA)	V_T = 18.6 MV
4-Ci source	(leakage current 12 μA)	V_T = 22.3 MV
6-Ci sources	(leakage current 19 μA)	V_T = 19.4 MV

The quoted voltages are thought to be accurate to $\sim \pm 1$ MV, due to uncertainty in the average column resistor values, limited accuracy in the generator voltmeter ratios and errors in the calibration of the neon flasher circuit measuring the dead-section and terminal current. One may conclude that the 6-Ci sources unbalance the column gradient to the extent that inner column breakdown sets in before the outer column is fully stressed.

Because a comparatively short time was spent at lower pressures, no good data have been obtained relating maximum voltage to gas pressure. Even at 110 psig (8.5 bar) we do not consider that the limiting voltage was achieved, since the tests were terminated before a voltage plateau was reached. Furthermore, the need to use a radiation source to obtain the highest voltages imposed an unequal gradient on the inner and outer columns. If this could be corrected by changing the column resistors or varying the charging conditions, it would be possible, in theory, to increase the gradient on the outer column by 25% to give a terminal voltage of $\sim$ 25 MV.

4. ANALYSIS OF THE TEST RESULTS

In this series of tests a number of foreign objects were removed from the column and progressive cleaning reduced the amount of debris in the tank. The SF_6 was dried so that most measurements were carried out with water concentrations below 20 p.p.m. The measured currents at the LE and HE column bases were always comparable and close in value to the current leaving the #3/#4 or #7/#8 dead-sections. The column current leaving the terminal, measured by neon flasher, was equal to or slightly greater than the column currents flowing into the #3/#4 or #7/#8 dead-section. The calculated increase in gradient due to this excess current in the inner column agreed to within the accuracy of measurement with the increased voltage measured by the central generator voltmeter relative to that at the portico dead-section. Thus the electrostatic conditions appeared to be stable and predictable in the presence of ionization currents almost equal to the base column currents. The effect of these ionization currents will be reduced when the tube resistors are installed, doubling the current drain down the column.

Polaroid photographs of most of the 160 sparks incurred during testing provided information on the nature and location of the discharges. In most cases an intense primary discharge from the terminal or the rings near the terminal to the portico is accompanied by several forked discharges from highly stressed areas of the portico to the tank (fig. 6). The tendency to spark downwards decreased after successive cleaning operations. At the highest voltages, a very large number of evenly distributed discharges appear to develop from the portico, suggesting that there is significant dispersion of the energy release.

5. CONCLUSIONS

The portico tests have demonstrated that significantly higher voltages can be achieved with the portico than without it. The maximum voltage obtained so far, V_T=22 MV, is consistent with the design aim and with the probable limitations of the accelerator tubes. Since it was achieved with a non-uniform column gradient, a further increase of ~2.5 MV could be expected when the column gradient can be made uniform. The portico appears to dissipate stored energy in a beneficial way, and no damage to the column or structure was observed in the course of the test. The radiation sources have an important effect on voltage, at least in the absence of accelerator tubes, and further improvements in performance may be gained by varying their strength, location and screening.

Two schemes for controlling column gradient are being considered. The resistors in the inner column could be reduced in value and a variable corona load drawn from the portico at dead-section #3/#4 or #7/#8. A better solution is to modify one of the Pelletron chains so that a controlled charge can be transferred from the base to the #3/#4 or #7/#8 dead-section, exactly compensating the additional current in the inner column. This requires the installation of a special take-off in the dead-section and separate inductor and suppressor power supplies for this chain.

This work has been carried out under a contract from the Department of Energy (DE-AC02-84ER40123) whose support is gratefully acknowledged. The tests reflect the efforts of the technical and operations staff and the enthusiastic participation of the graduate students. We are grateful to them and to T. Barker and R. Wagner for the new instrumentation.

REFERENCES

1) Hyder, H.R.McK., Baris, J., Gingell, C.E.L., McKay, J., Parker, P.D.MacD., Bromley, D.A. **Seventh Tandem Conference**, Berlin April 1987, to be published.

2) Letournel, M., **Proc. 3rd International Conf. on Electrostatic Accelerator Technology**, Oak Ridge (IEEE, New York, 1981).

FIGURE CAPTIONS

Figure 1.   Horizontal section through the ESTU tank, showing column, portico, dead-sections and "K" column transverse stiffeners. (Note: the portico panels do not actually lie on this section).

Figure 2.   Cross section through the center of the ESTU tank, showing column, portico, radiation sources, corona probe, generator voltmeter and capacitative pickups.

Figure 3.   Radial macroscopic field distributions in the ESTU with and without the portico, at 20 MV.

Figure 4.   Current leakage paths for ionization due to radiation from sources or bremsstrahlung.

Figure 5.   (a) Variation of leakage currents with radiation source strength at constant terminal voltage.
(b) Variation of portico leakage current with terminal voltage (2-Ci source).

Figure 6.   Spark breakdown at 21 MV. The primary breakdown, from terminal to portico, is obscured by the K-column in the foreground. Many secondary breakdowns between portico and tank are clearly visible.

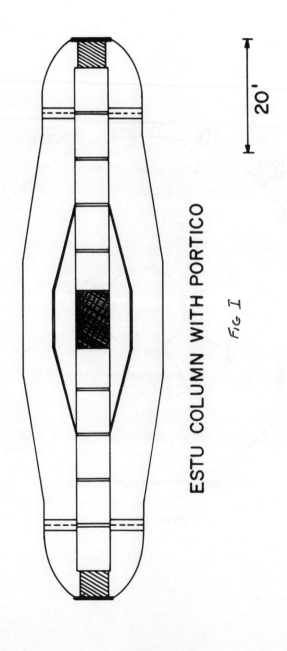

ESTU COLUMN WITH PORTICO

Fig I

20'

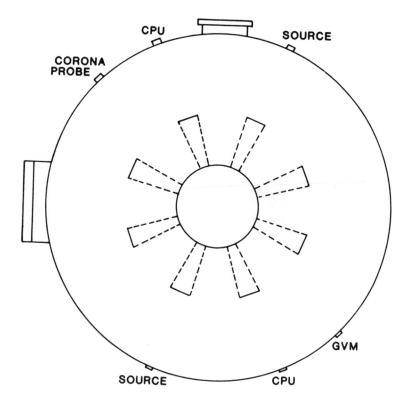

FiG 2.

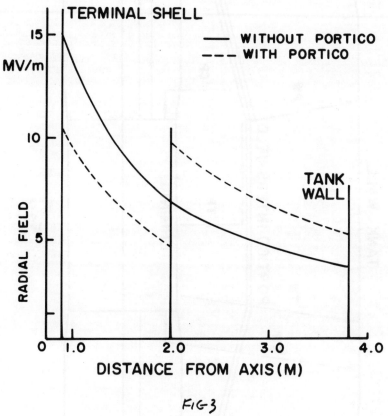

FIG-3

68

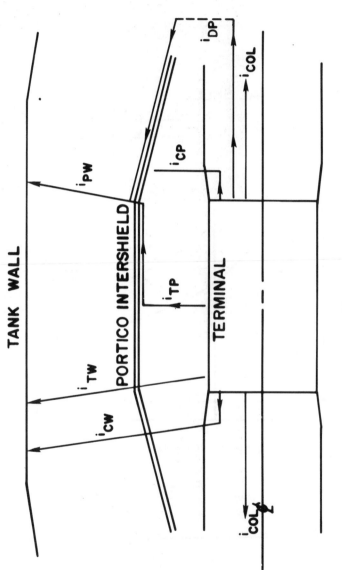

FIG 4

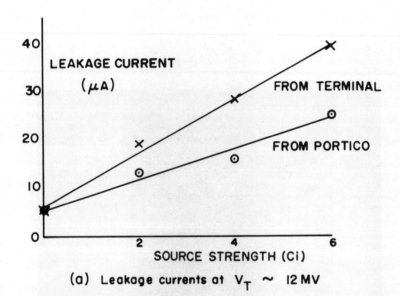

(a) Leakage currents at $V_T \sim 12\,\text{MV}$

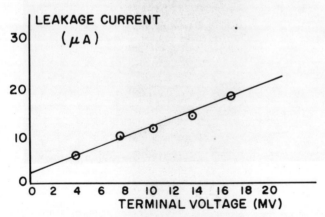

(b) Portico leakage currents with a 2-Ci source

FIG 5

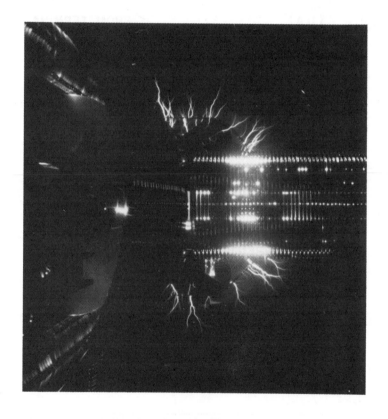

FIG 6

DISCUSSION

SCHMIDT: Does anyone have any questions?

STORM: University of Washington.   Do you have a schedule for completing the project, and do you want to talk about it?

HYDER: We hope to have the tubes installed in something like six weeks.   We then have to establish vacuum integrity through the machine and to install a temporary terminal assembly.

We have a lot of work to do on the control system.   We hope to have the beam early next year.   I do not think we will get it this year.

At some point after that, that will obviously depend on the experimental program, we will want to replace the initial terminal assembly with something that looks more like this.   This shows a short wide diameter gas stripper outlined in red, with two differential pumping canals also in red at either end of it.

By using a recirculating turbo molecular pump one can allow the diameter of the gas stripper to increase significantly above what is practicable for an uncirculated stripper, and we plan to put to foil strippers in between the differential pumping tubes and the gas stripper itself.

Our intention is to modify or have modified the standard NEC strippers in the way in which Ken Chapman described a year ago, so that each of them will hold something like 400 foils, so we will have a total capacity of about 800 foils.

This design is a compromise.   Following the gas stripper assembly there will be an offset quadrupole triplet charge selector designed by Off axis Dan the separator man and followed by a quadrupole singlet, a small element which will do a little focusing to complement the effect of the triplet.

We have 10 feet available from tube to tube to fit all this lot in.   The gas stripper can be made fairly short, the central part is half a meter long, but by the time one has added the foil strippers and the different pumping tube, the whole assembly is just about one meter long.

Of course, you get major benefit from recirculation if you fit the whole of this lot into a shroud which is pumped by the turbo and allow the outer ends of the differential pumping canal to be pumped by the 500 liter/second ion pump which pumps the rest of the terminal.

By providing a much higher conductance back to the ion pump than the conductance down the tubes, one expects to get pressures in the tubes which are in the 10^{-7} Torr range.

The off axis separator works best if there is a significant distance between the point at which the beam is scattered by the stripper and the entrance to the first element of the triplet.

If the first foil stripper is used in conjunction with the off axis separator, then one has something approaching that condition. If one uses the gas stripper and the off axis separator, and I suppose that's a combination which we may be forced to use, if we are using very heavy ions and we just can't afford the wear and tear on the foils, then again there is a reasonable separation between the middle of the gas stripper and the off axis separator.

The second foil stripper is really too close for satisfactory operation, but the argument for putting a second foil stripper there is that by following the gas stripper with a foil stripper we can modify the effects of coulomb explosion in the cases where one is injecting molecular species. The thought is, if we are worried about coulomb explosion or we are operating with fairly low charge state ions where the separator is not so vital we'd use the second foil stripper; if we were running with heavier ions where we have to sort one beam out of 256 different beams emerging from the second stripper then we'll try and use the first foil stripper.

MC NAUGHT: McMaster. I assume you have something on the console that reads terminal voltage. If the generating voltmeter doesn't really read terminal voltage what drives that meter?

HYDER: The answer at the moment is that we do not have a meter which reads terminal voltage. We have a meter which reads the field at the tank wall. That's what most people have, of course. They label the meter generator voltmeter or terminal voltage, but what it should really be is field at the tank wall.

In the case of a machine with a portico this is not a very direct measurement of the quantity that you are interested in. When we come to implement the control system for the whole machine we will take into account the measurement that we have of the voltage between the terminal and the portico.

There are a number of alternatives there. One can telemeter out current measurements, and if there's reasonable current balance between the current entering the inner column and the current leaving the inner column, then you can apply Ohms law to that number and add that number to the generating voltmeter reading to get the terminal voltage.

One can put a generator voltmeter in the terminal looking at the portico, and I think we will do this. That of course measures the field on the surface of the terminal. That is fairly directly related to the voltage between the portico and the terminal. So a combination of the reading of that instrument and the generator voltmeter on the tank wall will again give us an indication of the terminal voltage.

There is another possibility. The portico is physically large enough to accommodate the pancake type generator voltmeter. One could consider packaging a generator voltmeter and putting it in the thickness of one of the portico elements. There is more work in that than in putting something in the terminal, but it has the advantage that the instrument is in a lower field. But I suspect that we will probably go for the instrument in the terminal since others have done that and it has worked well.

CHAPMAN: Florida State. If I understood you correctly, the generating voltmeter in the tank wall effectively doesn't see the potential on the terminal, and yet it is expected in the Vivitron that it will. Is the structure sufficiently different that this is so?

HYDER: No, I don't think it is expected in the Vivitron that it will. I think that's right. I think the same general arguments apply to Vivitron as to this.

LETOURNEL: Strasbourg. There is the same argument, and we do not know how to measure properly the voltage of the machine.

HYDER: You can put a neon flasher at 20 points in the machine.

Unfortunately the neons we have used are not that stable, so the calibration of these instruments varies by several percent. They are a useful crude indication, but not high precision devices.

STORM: University of Washington. I have a question about the off axis charge separation. Maybe I am missing something, but the off axis triplet bends all the different charge states off the original beam axis, and I don't see how you get them back on again. How does it work?

LARSON: I didn't attempt to describe in detail how that works. First let me say it does work.

Secondly, the quadrupole singlet elements are displaced by different amounts. If you consider the displacement of a conventional quadrupole, ignore the quarter quadrupole concept, a physical displacement sideways is the equivalent of superposing a dipole field and a quadrupole field.

So in effect the three elements constitute three dipoles, and you take the beam off axis, bring it parallel to the axis in the center of the central element, and then by a mirror symmetric arrangement put it back onto the axis after that bend.

You are quite right that if this did the same thing to every charge state it would have no advantage, but it doesn't. It turns out that one charge state can be brought back on axis in this way, but other charge states are dispersed, and it is as if one had a few degrees of dipole bend in the system.

That is, the elements are equivalent to a few degrees of a dipole.

STORM: The point is that each element has different displacement?

LARSON: That is necessary if you have only triplet elements.

In the structures that I was describing for Strasbourg and Yale the addition of other focusing elements in the system in fact makes it possible to have all three of the triplet elements displaced equally, so it is not a necessary condition. It is one way of operating it.

But if you want to think of the concept of the device, then three elements alone have to be operated in the way I described, with different amounts of displacement.

SESSION IV

NEW LABORATORY, NEW ACCELERATOR, NEW CHALLENGES

J. R. Tesmer[a], C. R. Evans[a], and M. G. Hollander[b]

Los Alamos National Laboratory

Los Alamos, New Mexico 87545

The Los Alamos Ion Beam Materials Laboratory (IBML) is a new, compact, user operated laboratory, dedicated to the characterization and modification of the near surfaces of materials. It is operated by the Condensed Matter and Thermal Physics Group of Los Alamos National Laboratory's, Physics Division and sponsored by the Center for Materials Science. The IBML is directed by a steering committee consisting of research team leaders and other interested parties. There are two accelerators housed in the laboratory: a 200 kV Varian CF 3000 ion implanter and a National Electrostatics Corp. (NEC) 3 MV tandem electrostatic accelerator. Because of the wide use and knowledge of the implanter, this report will deal primarily with the tandem.

1. CONSTRUCTION

Funding for the IBML was obtained in the summer of 1985 with the site for the IBML chosen to be in the basement of the Condensed Matter and Thermal Physics Group's building. This location offers installation challenges due to being served by only a four-ton elevator and having no overhead crane. Further difficulties were presented by the previous use of this area for storage and as the location of large

a) Physics Division
b) Mechanical and Electrical Engineering Division

mechanical vacuum pumps (some of which were still in use). These problems were solved by designing the tandem tank to split in halves and the pumps were either moved elsewhere or mounted near the ceiling.

Installation of the tandem proceeded smoothly with less than two months between delivery and operation for experiments (February 1987). There are always minor problems such as not having a flat floor for rolling the tandem between its operating position and its service position. It requires as many as six people to accomplish the move. In addition, the uneven floor places nonuniform stress on the accelerator's wheels -- we have destroyed two so far.

Site preparation and installation were accomplished by Los Alamos crafts, the research teams, and outside visitors. Total investment in the IBML, including much of the experimental stations, is ~$1.7 million -- funded out of Laboratory-wide resources.

2. OPERATION

The scheduling and operation of the IBML is designed to be extremely flexible to allow for maximum use and the rapid turn-around of experiments. Scheduling for the tandem is done weekly with users signing up for ~4hr time periods. Days not scheduled at this weekly meeting are available for scheduling anytime afterwards.

The users are the operators; they may start the accelerator whenever they are ready. Occasionally, users require longer periods of operation (e.g. around-the-clock). In these situations it is required that the user provide periods of ~8 hrs at least twice a week for other users short-term experiments.

This mode of operation is clearly the type necessary for most ion-beam characterization and modification experiments. Ease of operation of the accelerator is necessary to accomplish this.

3. PERFORMANCE

3.1 Implanter

The implanter uses a wide variety of ion source feed materials and in general operates very well. It is highly automated allowing for the average user to quickly become a reasonably skilled operator without undue hazards to either operator or machine. A major problem is in obtaining its rated voltage at the altitude of Los Alamos. The 200 kV can only be accomplished by injecting freon around its high-voltage terminal. Another major problem is ion-source maintenance and beam development. Although there has been much literature written on implanter ion beams there is still a great deal of black magic involved. No hazardous materials are used as source materials.

3.2   Tandem

The tandem is a new model from NEC. It was the second of this type built but the first put in operation. The accelerator is cryopumped (except for the alpha source) and uses all metal vacuum seals—mostly of the conflat type. Vacuum pressures are in the low 10^{-8} torr when not in operation. Although a foil stripping system was purchased, to date, only gas stripping has been used. This is the result of the very successful use of a recirculating turbopumped gas stripper.

The first question asked about a new tandem is, "Does it meet voltage specifications?" In this case the answer is an unqualified yes! Indeed this may actually be a 3.5 MV tandem (I don't know of anyone who has had the time or courage to try it). The terminal can be rapidly raised to the rated voltage with almost no conditioning. Some of the good voltage performance is probably due to NEC's departure from corona point voltage grading to the more widely used resistor system. Another reason may be the use of recently designed accelerating tubes.

Beam quality from this accelerator has been surprising. Although the emittance of the beams have been crudely measured, the alpha beam has the best emittance of ~1.2 π mm mrad(MeV)$^{1/2}$ for 80% of the beam -- a surprisingly low value! Hydrogen beams from the same source are around 3 π mm mrad(MeV)$^{1/2}$ while beams from the SNICS source are around 4 π mm mrad(MeV)$^{1/2}$. The difference in the quality between the alpha and proton beams may be due to the lower probe voltage for the proton beam and the inability of the injection magnet

to separate the different energy H⁻ beams produced from the positive
molecular beams from the rf source. Most of the beam used for experiments to date (well over 1000 hrs) has been alphas. The alphatross
source produced by NEC may indeed operate on black magic. In principle it is a very simple rf source followed by a rubidium charge
exchange region. Without great care (and some praying) the lifetime
of this source can be as short as 8 hrs. The mode of failure is
usually the shorting of the boron nitride insulator for the extractor
in the rf bottle. Rubidium contamination may play a major role in
this phenomena. NEC has made a design change which helps contain the
rubidium. However, we have found that the trick(s) to obtaining
longer lifetimes has been to leave the source running continuously
with source magnet and probe voltage at half value, oven and other
power supplies off. After experimenting with softer graphite and
tantalum, we now make the rf bottle extractor out of Poco AXM
graphite. We have found that tantalum works particularly poorly when
operating with hydrogen. Our lifetimes have averaged about 300 hrs.

The SNICS heavy-ion sputter source has seen very limited use.
However, it met all performance specifications once NEC determined
(believe it or not) that our altitude (~7500 feet) was affecting the
freon cooling system. Pressurizing the cooling system to ~10 psi
solved this problem.

Beam transmission is also relatively good if the accelerating
tube gradient at the entrance to the accelerator is always operated at
close to full gradient for almost any terminal voltage. Unfortunately, the column shorting system is relatively hard to use and its
use shorts out the high energy column current metering.

4.   PERFORMANCE EFFECTS ON EXPERIMENTS

Most of our materials characterization research revolves around
either Rutherford backscattering (RBS) analysis or nuclear resonance
reactions. Both of these techniques are very sensitive to the reproducible energy of the accelerator day after day. This presents a
large challenge to the tandem accelerator system because of its
compact design. To keep the length of the accelerator short the

analyzing magnet was designed without a well-defined object to image. Therefore, changes in tuning can effect the beam energy by several keV. This is not extremely serious for RBS data but is a problem for resonance reactions. The only solution so far achieved has been careful tuning and blind faith in the accuracy of the generating voltmeter readout. In fact the generating voltmeter has been extremely reliable providing indications of energy variation problems caused by tuning.

The terminal voltage stabilization system has also presented problems due to its oscillation when there is small amounts of beam on the regulating slits or when the slits are close together. Another problem is the inability of the system to center the beam between the slits without human intervention.

5.   SUMMARY

The IBML has rapidly become a valuable Los Alamos resource. This was largely due to the rapid installation, reliable operation, and performance of the tandem. The accelerator is not without problems, but its versatility and design have allowed for almost instant utilization in a variety of experiments at a time of excitment in ion-beam materials research.

The IBML falls into the Department of Energy (DOE) category of user resources. The IBML's primary purpose is to serve the needs of in-house laboratory staff, but may be available to DOE-supported universities and other qualified users subject to local management scheduling decisions and other applicable restraints.

SPECIFICATIONS

Terminal Voltage: 3 MV +

Charging Current: 300 μA

Beam Specifications:

H+	1 MeV	2 μA
H+	3 MeV	3 μA
He++	4 MeV	2 μA
He++	9 MeV	1 μA
B+++	12 MeV	12 μA
N+++	9 MeV	0.4 μA
Si++	9 MeV	20 μA

BEAM QUALITY
(Accelerated Beams)
(Horizontal)

BEAM	EMITTANCE (80%) (Π mm mrad MeV1/2)
H+	3
He+	1
Fe++	4

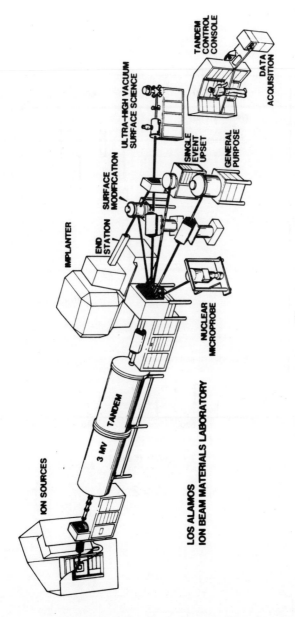

ION SOURCES

3 MV TANDEM

IMPLANTER

END STATION

SURFACE MODIFICATION

ULTRA-HIGH VACUUM SURFACE SCIENCE

SINGLE EVENT UPSET

GENERAL PURPOSE

TANDEM CONTROL CONSOLE

DATA ACQUISITION

NUCLEAR MICROPROBE

LOS ALAMOS ION BEAM MATERIALS LABORATORY

FIG 1

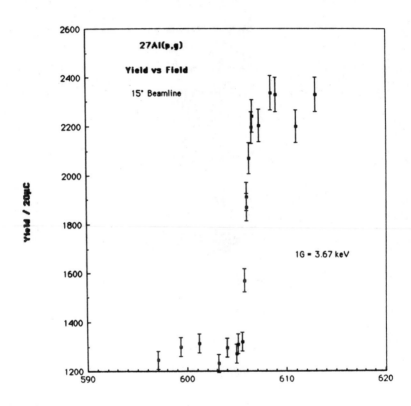

27Al(p,g)

Yield vs Field

15° Beamline

1G = 3.67 keV

Yield / 20µC

Field (Gauss)

FiG 2

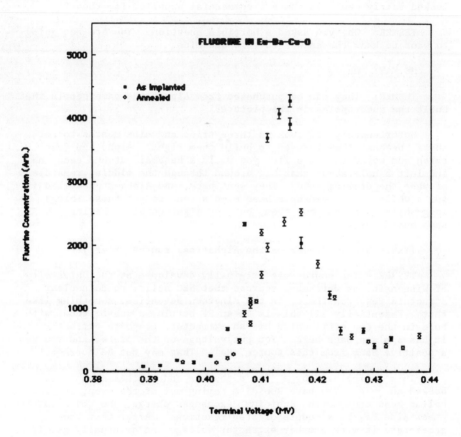

FIG 3

Discussion Following J. Tesmer Paper

JANSEN: Are there any questions?

MC KAY: I notice you were using box magnet type steerers that looked fairly new. Is there a commercial supplier for those?

TESMER: Oh, you asked a horrible question. You are not going to want to know the answer to this question.

MC KAY: Oh.

TESMER: They can be purchased from Alpha, the same people that built the quadrapoles in the picture.

Unfortunately, it took us three tries and nine months to get those, because they refused to build them right. Alpha had just taken and wound the wire like you would a baseball around each one and left a hole about that big around through the middle, regardless of what the drawing said. They went back, and Alpha stonewalled it for a while. It could have been even a year to get those things. A long while. You can buy them, but you might consider having your shop build them.

IYER: Can you describe the Alphatross source briefly?

TESMER: The source was originally developed at the University of Wisconsin, as were most sources that NEC sells, to do nuclear structure research there, in the nineteen seventies, something like that. Essentially all this is is an RF bottle, a quartz tube, with a hole in the end of it which has an extractor, graphite extractor, and it has a plate back here. You put voltage on the plate, and you get a positive beam from this source, He^+. That may not be correct.

This is followed by a "T", a chamber and to this chamber is hooked an oven. You have Rubidium coming out of the oven, and the helium beam crosses the Rubidium, exchanges charge, and comes out the other side He- by a couple of step reactions. After that you accelerate it with another extractor voltage and eventually get it up to 50 or 60 kilovolts. Does that explain it?

IYER: Yes.

TESMER: I can talk to you some about that later. I have become an expert.

MC KAY: How big is the aperture in the RF bottle?

TESMER: Number 40 drill. It is about 1/16th of an inch diameter and about 1/2" long, it's a canal essentially.

ALPHATROSS ION SOURCE

The source is patterned after the efficient RF positive ion source - rubidium metal charge exchange negative ion source built and in regular use by Professor H.T. Richards at the University of Wisconsin-Madison, Department of Physics for about the past six years.

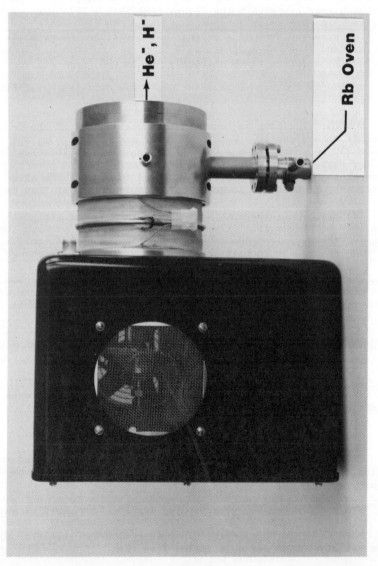

MC KAY: That's different from an RF source.

WEIL: It's a hole in a piece of graphite? The canal material is graphite?

TESMER: Yes. A magnification of this looks something like this. Here's a cross-section. It is inserted in something that looks very similar to it, made out of boron nitrite, so it has another bigger piece with the flange like this. This part is boron nitrite. It then seals altogether with an "O" ring, which is here, and it presses up against a recess that positions the graphite aperture and then the bottle itself is here. So this all is here together by a spring clamp mechanism on the end of it.

WEIL: Is that canal good for high beam currents, a milliamp or so?

NORTON: You have to have about a milliamp to get the beam currents we are seeing. We did an experiment a few years ago and had around 950 μA. Most of our work is taking place in this region. The kind of carbon used is critical. We recently changed to a high temperature "O" ring from a viton one and it worked. So with this source the old black magic seems to be concentrated at this point.

TESMER: What causes this source to fail is not the bottle. It's easy to change it. If you let it up to argon, you pop the "O" ring and these two components out and stick another assembly in, you are back on the air in half an hour. It's very easy to change.

What happens is the boron nitride gets coated with carbon, and when it does that it shorts out the plasma. All the voltage drop is occurring right here. Even though the bottle is seven inches long, the voltage drop is really right up in the front of it.

That's where you get the accelerating potential. When you short it out, the source looks like it's doing great, and nothing comes out.

JANSEN: Yes, we have the same experience. We don't make it negative, we leave it positive.

ADAMS: Is that an RF source? It doesn't show a coil.

TESMER: It has two electrodes on here, for the RF, and in this region there is a magnet to press the plasma down. A very necessary item.

STORM: Did you buy your bending magnets from Alpha or who?

TESMER: From Magna Coil. They came with the accelerator. We were not allowed to buy quadrapoles from them.

STORM: Do you want to explain that or not?

TESMER: Well, I guess I could explain it.

Magna Coil was the lowest bidder on quadrapoles, but Los Alamos has a lawsuit with them, which is still ongoing.

They made some magnet for LAMPF and it didn't work, and they couldn't reach a settlement.   It has gone into litigation.

MAN IN AUDIENCE: Did they make the bending magnet for NEC that they sold to you.

TESMER: That's right.   The magnet was quite good that Magna Coil built.   There are some problems with the magnet chamber which Magna Coil also built.   It is very complicated, because it has all these ports coming out, and when you go to weld the long tube on you will not get it straight.

So the alignment ports are almost useless.   They don't point in any direction where the beam might come out.   The bending magnet is really quite good.   The NMR probes work very well so the field must be relatively uniform.

MAN IN AUDIENCE: Livermore.   Is the material the extractor made out of on the RF source critical?

TESMER: I think it is.   Tantalum does not work.   Do not use tantalum except for, it works for alphas but it is terrible for hydrogen.   It will kill that.   We are using something called AXM, graphite made by Poco.

There are lots of grades of graphite.   AXM is the hardest grade, I think.

MAN IN AUDIENCE: Is that what gives the helium beam?   Aluminum will work for hydrogen.

TESMER: I haven't tried it.   NEC has been doing quite a bit of research into it.

NORTON: We used aluminum for some time and got poorer lifetimes.   Sometime ago Poco changed the formula and didn't tell us. We got almost no beam out of the source.   We went back to them and tried different grades and came up with this, and it seems to work until they change the formulation again.

This is where most of our work is going with the sources in this area.

JANSEN: I think we used a stainless steel canal.

TESMER: Stainless steel doesn't work for hydrogen.

JANSEN: It works for us.

WEIL: WE are getting a milliamp for three months, two months, out of stainless steel canals. But the stainless steel is eroding in one or two weeks running time at 1 milliamp. I would like to ask what is the typical lifetime of these graphite sleeves.

NORTON: Pretty much what we hear from Los Alamos and Texas. They have had one for four years. They run 300 hours, scheduled shut down, and change the RF bottle. What we are seeing is a mode of operation where you keep the source sort of on, the low voltage turned down, gas running in, 300 hours, is very straightforward.

TESMER: You leave the source on or the rubidium will get back and attack the graphite. It will migrate right through the graphite which will swell. This does not happen with tantalum but the rubidium will still affect the boron nitrite. You get very stable α beams with tantalum. There is a sputtering problem with tantalum.

JANSEN: Oh, yes, there is, and as you said it will short out eventually.

GRAY: Does the alphatross have a quartz sleeve?

JANSEN: Well, I think the boron nitrite takes its place.

SOUTHON: The number you gave for the acceptance of the stripper canal. What is the transmission of this? Do you get everything through?

TESMER: It depends on the energy you are running. It doesn't have a gridded entrance lens, if you can get the entrance lens up, the gradient up, to where it should be, you get something like I think by memory 25 or 30 per cent with a gas stripper.

We find a gas stripper for our purposes, we don't need much beam at this stage of the game, to be quite adequate. If we have a foil stripper, this wasn't built for the heavy ions.

JANSEN: Thank you, Joe.

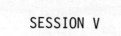

SESSION V

TUTORIAL ON COMPUTER CONTROL*

R. C. Juras
Oak Ridge National Laboratory
Oak Ridge, Tennessee 37831

Thank you. The word tutorial in the title for this talk is probably a little ambitious. Ken asked me to talk about computer control and maybe stimulate some discussion.

I should mention my experience with computer control. I have worked with the Oak Ridge Holifield Tandem facility control system since 1976. As most of you know, that's a totally computer based system. Before that I worked at Sperry on a computer control system for an aircraft that was to mimic the characteristics of the space shuttle during landing. It is used for astronaut training.

What comes first? First you must decide if you want a computer-based control system. Tandem accelerators, particularly smaller ones, can run just fine without computer-based control systems. They've been doing it for years, successfully putting beams on target. However, computer based controls offer advantages, particularly for large tandem accelerators.

What are the disadvantages of computer control? There are several disadvantages. Complexity is a disadvantage. You have to write software and people have to understand how it works and how to keep it running.

You have to write custom software because, although you can buy a wide variety of software these days, you can't buy much that's applicable to accelerator control. As the accelerator evolves, new controls must be integrated and software rewritten. So, you must spend time maintaining the accelerator control software.

You need to do system management. You have to do backups and

*Research sponsored by the Division of Basic Energy Sciences, U.S. Department of Energy, under contract DE-AC05-84OR21400 with Martin Marietta Energy Systems, Inc.

This manuscript is an edited version of the author's remarks. This format has been chosen to facilitate timely publication.

worry about maintaining the operating system. When the computer
vendor upgrades the operating system, for example, some of your
programs may have to be changed because the details of system calls
have changed. These sorts of things are a drain on manpower.

It can be difficult to integrate computer-based controls with
some older equipment. Many of the older power supplies were not
built with the intention of remote control or computer-based
control. It takes some effort and expense to modify or replace
those supplies.

Once you have a computer-based control system, accelerator
operation is dependent on the hardware and software reliability.
In fact, the computer system can even damage the accelerator if the
software does crazy things as a result of bugs. And since the
number of hours of beam on target is dependent on the reliability,
you have to worry about making it work right and getting all the
bugs out.

Given all those disadvantages, why have computer control? I
believe the principal benefit is software aids-to-operation. Once
the computer-based control system is in place, you're limited only
by your imagination as to what you can do with it.

You can -- if you have the resources -- write completely
automatic controls for accelerators. I believe that accelerators
can be tuned automatically, but I don't think we'll see that soon
because it's too expensive given the present state of software
engineering.

But there are types of software which can be provided with
much less cost that provide great benefit for accelerator operation
--software, for example, to cycle analyzing magnets automatically
(to avoid hysteresis), to set up the accelerator from scaled,
recorded values from previous runs, to scan the ion source
analyzing magnet and plot the results, or to keep a running log of
accelerator parameters.

Controls can be easily multiplexed for operation at elevated
potentials and this can be done at little extra expense because
computer-based control systems multiplex data on serial data links

anyway. Of course, you have to worry about spark protection of the electronics at elevated potential, but it has been demonstrated, in a number of applications, that this is a solvable problem.

Computer-control systems offer precise, repeatable controls so that you can go back to exactly where you were at some previous time. Elimination of long cables and ground loops is advantageous, particularly for larger tandems. It is expensive to pull many long cables throughout a building and the resulting ground loops can be a difficult battle to fight. Computer systems operate with just a few serial data link cables between equipment racks and these serial links are easily ground isolated.

Record keeping is another advantage. You can keep detailed records of all the beams you have run with little effort.

New controls are usually easy to add although this can sometimes be difficult. The wiring is always easy. There are no long cables to pull, just a cable to the nearest equipment rack. The software is easy if it's another device of the same type that you already have. The software is difficult if it's a brand new type of device so that you have to write a whole section of new software and update every existing program.

The remainder of this talk will be divided into three parts, the three important components of control systems: hardware, software and the operator interface.

HARDWARE

What considerations should be kept in mind for hardware selection? Select computer hardware so as to reduce software costs. Why do I say that? Most control hardware can be purchased these days and the price performance ratio is improving rapidly. It doesn't take a lot of manpower to buy computer hardware. Just buy it and bring it in. It can be done quickly. But almost all the accelerator control software must be custom designed. So, I think it's necessary to keep software cost in mind when deciding on hardware. Spending a little bit more on hardware may reduce the total system cost.

You may buy a cheap computer only to find that you have to write tens of thousands of dollars of software to get the system to work right because, for example, the software must be written in assembly language or some other low-level language to achieve the execution speed you require.

Select a computer system that can be upgraded without rewriting software. We all know that computers rapidly become obsolete. They're obsolete sometimes before you get them in. So, someday you're going to want to upgrade your computer. The problem is the software. If the software's very computer dependent, it may lock you into an obsolete computer. Assembly language software, for example, is computer-dependent. If the software is written in a high level language, it is more portable, but even then you're probably going to have to rewrite some of the I/O interface.

If your computer is a member of a compatible family of computers, even assembly language software can be portable. Examples of this are the VAX or IBM PC families. In such a family, there exists a range of computers from the low end to the high end. So you can start with an inexpensive, low end computer and move up to more powerful computers keeping the same software all along the line including, in many cases, the I/O software.

We are using Concurrent computers (Concurrent is a subsidiary of Perkin-Elmer). Concurrent turned out to be a good choice for us. We were able to replace the original computers and install more powerful computers with minimal software changes.

Another way to facilitate upgrades is to use computer independent buses and I/O. CAMAC, for example, is independent of the computer that you use. When you upgrade your computer you don't have to change any of your A/D converters, D/A converters, or other input or output modules. The accelerator interface stays the same. Avoid A/D and D/A convertor systems that are tied to whatever computer bus you are using unless, of course, your computer bus is itself a standard, such as VME.

Another way to save on total system cost is to standardize the hardware interface. This is a little harder to define. I'll give an example shortly.

If you can standardize the hardware, the software will cost less to write in the first place because you will be able to make use of common subroutines and common procedures. In other words, if accelerator interfaces are identical, or can be made to look identical to the software, common software can be used.

Hardware upgrades that conform to the standard can be accommodated by simple database additions rather than new software. And the software is easier to maintain. There is less of it. It is easier to understand.

Here is an example (Fig. 1). All of the devices, A, B and C, are the same type. They all require the same sort of conversion, maybe a linear conversion, by the software. However, the details of the interface to the three devices differ slightly from one to another. The three rectangular boxes represent the binary number read by the computer from each of the three digital input registers (CAMAC modules) that are wired to the devices.

For devices A and C, logic 0 is a fault condition. Logic 1 is a fault for device B -- probably the interface module was simply wired to the other side of a form C relay contact. The other differences are simply due to differences in the wiring to the interface module.

What is the result of these small differences? The software shown on the left of the viewgraph is necessary to deal with the three devices wired as shown. If all the devices were wired like device B, the software on the right would result.

This, of course, is a trivial example. However, there are real-life examples that don't differ much. At the Workshop on Accelerator Controls, Michael Glass[1] talked about experiences at Fermilab where quite a bit of involved software was written to make up for the lack of an inexpensive integrated circuit in a CAMAC module.

Here are some rules of thumb for interfaces. Try to use common components whenever possible. Why? Because there will be fewer spare parts to keep in your inventory and there will be less special software to write. I'm talking here of A/D, D/A and other interface modules.

Provide true device status to the operator whenever possible. When the operator pushes a button on the console to turn on a power supply, for example, use a relay in the power supply to return the true status instead of just echoing his request to turn on the supply as the power supply "on" status.

Leave some room for growth. It doesn't cost much to leave spare wires in cables; you're probably going to need them later. Leave a little extra panel space, too.

Every interlocked device should have an indicator signal for each interlock instead of one indicator signal for several interlocks that are in series and perhaps scattered throughout the building. This implies that you should have a pair of wires coming to the interface from each interlock. Beam time is expensive. Without individual indicators, if you have a fault that is keeping beam off target you may have to rush through the facility to find which interlock is the culprit.

SOFTWARE

We've already talked a little about software because the hardware and software are interrelated. Here are a few other ideas about software.

Additions and changes to the software are easier and faster if you use a database. For one thing, with a database there is no need to work in a cryptic format. Instead you work with easily understood text format with powerful text editors. That means that there is a smaller probability of error and that errors are easier to spot; you don't have to look at something encoded in hexadecimal, but can quickly scan through unencoded text.

Complex interrelations in the data buffer can be changed automatically by the database program. The cross-referencing is done every time a run-time buffer is generated.

The documentation is improved. Listings can be made with the data sorted in various ways. Editors can be used to search the database, for example, for every occurrence of a particular type of device or every device turned on and off by a particular I/O module. Application programs can access the database for up-to-date information rather than relying on conversion factors and addresses buried within the application programs.

Figure 2 is an example of the value of a database. In this example it is seen that the database is easy to read. The database program is used to translate the accelerator database into the buffer image format dictated by the computer hardware. Imagine being faced with the need to make changes to the buffer image, perhaps because a power supply on the beam line has been replaced. Contrast the ease of changing the database with the difficulty of changing the buffer directly. In addition, the database program can make listings for documentation.

Figure 3 illustrates another value of a database. In this case, the value X is read from an accelerator component. It is typically a number read from an analog-to-digital converter. Several programs may access this number. Each program then must convert this number to a meaningful value perhaps by multiplying by a constant, A. The number X may be a reading from a rotating coil gaussmeter, for example. If the rotating coil gaussmeter is serviced and replaced, the constant, A, may change. Then you are faced with changing the constant in each of the programs in which it occurs. First you must determine which programs contain the constant, then find the source code. If the programs are written in a compiled high-level language, such as FORTRAN, they will have to be recompiled. On the other hand, if the constant is stored in a database that the programs can access, making the change is simply a matter of replacing the constant in the one place it occurs.

Anytime you can, use a high-level language. The use of a high-level language increases programmer productivity. Because it is easier to understand and modify, it reduces maintenance costs.

In addition, high-level languages are more transportable to new computers.

Not so long ago it was necessary to write time-critical software in assembly language or FORTRAN with in-line assembly code. The FORTRAN, and other high-level-language, compilers of ten years ago were not very efficient at generating good machine language code, a code that is compact and executes very quickly.

Compiler technology has come a long way. There are now 'optimizing' compilers available for FORTRAN and other high-level languages that produce machine code as good as that of a proficient assembly language programmer. Typically, you buy two compilers. One compiler is used for development of your programs. It compiles quickly but doesn't result in optimal machine code. You use this compiler until your program is debugged. Once the errors are out of your program, you compile the program with the optimizing compiler. It takes a much longer time to compile, but it results in compact code that executes more quickly.

Provide a good programming environment. A good programming environment leads to better application programs. By "good programming environment" I mean a powerful operating system and transparent access to real time accelerator data. The people who design the computer hardware and system software should try to make it easy for those who are going to come in later to write application programs.

In our case, the scheme that is used to provide access to real-time accelerator data is a shared memory between the computer that runs the application programs and the computer that runs the accelerator. All parameters are kept in the shared memory by the accelerator control computer. The application programmers don't have to worry about scheduling I/O or requesting data from the control computer. Application programs simply read values from shared memory; the data is never more than 1/2 second old. Or the application program writes to shared memory; the data will be sent to the controlled device within 1/2 second.

It is also important to provide the application programmer access to the accelerator parameter database. Don't use the database for just your systems programs. Make it available for application programs.

A friendly operating system, a high level language and conceptually simple access to data make programming very straightforward. The result is that staff members, visitors on short-term assignment, graduate students and co-op students can write application programs.

Having short-term people writing software can be a double-edged sword. They may produce good software, but when they leave you must maintain the software. This is another reason to use a high-level language.

OPERATOR INTERFACE

The final topic to be discussed is the operator interface. This is the part that is likely to stimulate the most discussion. A computer-based control system represents an opportunity to create a better accelerator control environment -- or a worse one, if not done carefully. The operator interface can have a big effect on efficiency of operation.

The operator interface is important also because a significant portion of the control system software must be written for the operator interface.

Let me quickly describe our system. Then I'll try to tell you some things that we have found that are wrong with our system.

We have two identical control consoles. Each console has a color alphanumeric display with an associated page selector and trackball driven cursor. The cursor is moved to an item of interest, for example the on/off control for a device, and a button located next to the trackball is pressed to change the status of the device.

For those who may not know, a trackball is simply a mouse turned upside-down and mounted on a tabletop. If you buy an upside-down mouse for the Macintosh computer, it is called a turbomouse. So I guess we have a turbomouse.

Each console has three 'assignable' shaft encoders. The cursor is moved to a CRT line, the 'assign' button next to a particular shaft encoder is pressed and the shaft encoder then controls the device pointed to by the cursor.

Three analog meters on each console are assigned in a similar manner. Analog meters are useful for trending information, for optimizing the beam on a faraday cup, for example. Eight 'analog jacks' can be assigned, also. The output of each analog jack is a zero to ten volt signal that is typically connected to a strip chart recorder and is used to monitor such things as vacuum in dead sections of the accelerator, the GVM and x-ray levels.

In addition to the assignable components, there are six dedicated meters for terminal potential stabilizer parameters. Oscilloscopes are used to display NMR resonances, beam profiles and the terminal capacitive pick-up. The oscilloscope signals do not go through the computers. The NMR is set by the computer, but the resonance condition is detected by the operator looking at the NMR oscilloscope. There are several emergency shutdown buttons. They are comfort buttons that are independent of the computer. The computer has been exceptionally reliable so far and we haven't had much use for them, except during maintenance.

Now I will detail the shaft encoder operation. When the shaft encoder is assigned, the device label appears above the shaft encoder. This is a valuable means for the operator to keep track of which shaft encoder is assigned to which device.

'SAVE' and 'RESTORE' buttons are provided for each shaft encoder. Typically, when the operator assigns a shaft encoder, he presses the SAVE button. Then if the result of his tuning is a decrease in beam current, he presses the RESTORE and the beam current returns to its previous value. This really works and is used often. This is one benefit of the precise, repeatable nature of digital controls.

Each shaft encoder has a 12-position range switch for two to 4096 turns full scale. Thus the operator has a choice of the resolution of the shaft encoder. Percent of full scale output for the assigned device is displayed on the CRT.

About a year ago the method of shaft encoder assignment was changed slightly. Previously, if the assign button was pressed with the cursor on a CRT line with no legal assignment, the knob assignment was nulled. Now the previous shaft encoder assignment is retrieved. So it's possible to toggle back and forth quickly between two assignments without reference to the CRT page.

After five years of experience, we're still happy with the use of the shaft encoders and the philosophy of multiplexed control, but agree with the operators that more knobs are needed. The consoles have three shaft encoders, but many controls naturally occur in pairs of pairs. An example is the x and y axis of a quadrupole triplet followed by the x and y axis of a steerer. So, the number of shaft encoders per console should be increased to at least four. Five is the number we're probably going to settle on. There doesn't appear to be a need for more than that.

The shaft encoders are now read and sent to the controlled device 50 times a second except for the bending magnets. The 20 bit digital-to-analog converter for the bending magnets is an electromechanical device and to avoid wear it is updated only twice per second. Our experience with bending magnet control leads us to believe that maybe twice a second is fast enough for all control devices. So, we may try changing all the shaft encoders to a 2 Hz update.

Analog meter assignment is just like the shaft encoder assignment. The device label appears above the meter, and the range and range units appear below the meter. Each meter has a times three switch that expands the lower one third of the meter to the full meter scale. The meters are updated 50 times per second.

Following the success of the change in the method of shaft encoder assignment to recall the previous assignment, the CRT page selection was modified. Pressing the Enter key on the key pad with no number entered now recalls the previous page. It is possible to toggle between two CRT pages simply by pressing ENTER.

The reason I stress these modifications is that they illustrate a valuable point: In a computer control system, the ability to go back one step is beneficial. Even the save and restore buttons on the shaft encoders can be considered an implementation of this general philosophy. You should provide the operator with a means to go back one step wherever possible.

Another thing we have found since our control system was built is that alarm indications are necessary. In control systems with multiple CRT pages, only one CRT page is visible at a given time. The operator must be made aware of problems on other pages. Our present solution is a bank of alarm lights. Each alarm light has a legend indicating either a particular alarm condition (such as a serial highway error) or a CRT page number. If an alarm indicator with a CRT page number lights, the operator turns to that CRT page. On the page, the alarm is highlighted by a red background. The alarm may be acknowledged by moving the cursor to the alarm. The background for that alarm then turns yellow and the light on the panel extinguishes so that the operator will be aware of another alarm occurring on the page. When the alarm condition goes away, the alarm is rearmed.

We are looking at a new way to present alarms, perhaps a small CRT display, so the operator doesn't have to change pages. Alarms are an important part of a control system and require careful thought.

Software aids to operation are possibly the major benefit of a computer-based control system. I'd like to simply list here some of the generic things you can do to aid the operators.

You can write software to record beam parameters for successful beam tunings. Then, in the future, you can set up the accelerator for new beams based on scaling the recorded parameters.

Software can identify ion species. The computer can scan the mass analyzing magnet following the ion source to determine the ion source output, much like a mass spectrometer. Programs can also be written to look at the accelerator terminal GVM and magnet NMR's to verify which ion species is coming through the accelerator.

Software can log accelerator parameters both at timed intervals and on demand. The operator may then look back at the accelerator history. For example we recently discovered degraded vacuum in an external accelerator beam line. The question then was: How fast is the vacuum degrading? Must we shut down now or can we wait for a break in the schedule? In this case, the history log showed it had been getting worse slowly for quite a while, so repairs could wait.

Analyzing magnets can be cycled automatically to eliminate effects of hysteresis. This used to be a tedious chore for the operators. Programs can be written to calculate theoretical foil stripper lifetimes and to calculate parameters for manual setup (including charge state fractions after the strippers).

I now want to discuss a subject which is almost sure to stimulate discussion, alternative operator interfaces. As the technology changes, we should think about ways to utilize the new technology to improve the efficiency of accelerator operation.

An alternative to the trackball or mouse controls described previously is a touch panel display with menus. The menu systems are easier to learn in the first place, but the operators I've talked to complain that in many such systems, even experienced users must go through the tree structured menu for each operation. If the tree structure is several layers deep, the operator will soon become frustrated with the time required.

Our system, with almost 100 pages, for one example, is a little harder for the operators to learn in the first place, but after they become experienced they can remember which page almost every device is on. The capacity for recall in humans is great.

Which interface is best depends on the situation. Menus are great for inexperienced or infrequent users while full-time operators probably prefer a more direct interface.

Because of less expensive graphics along with software to help you do graphics, high resolution graphics displays might be the future. Expect to see more McIntosh-type displays with a mouse (or trackball) and pull-down or pop-up menus. The Vivitron designers are considering doing their operator interface with pull-down menus and I'm very interested to see how that turns out. It may be very good. Designers at Fermilab are also working on an operator interface using that paradigm.

Up/down buttons versus shaft encoders are a continuing debate. We have control systems that use each. The cyclotron control system makes use of buttons while the tandem control system makes use of shaft encoders. When we began planning our upgrade, we asked the operators if they preferred up/down buttons or shaft encoders and they chose shaft encoders. One advantage of the shaft encoders, they said, was without looking you can go up two turns and down two turns and return to where you started.

Alternative analog displays are available. Bar-graph displays could be used in place of analog meters, particularly now that high-resolution displays are available. Another alternative is a meter or bar-graph display that is simulated on a graphics CRT display. I believe that the Vivitron designers are considering the use of meters simulated on a graphics CRT display.

CONCLUSION

In conclusion, a good operator interface is important for efficient accelerator utilization. The operator interface is going to require a lot of software so it should be well thought out at the beginning.

The hardware for computer-based control is readily available now. It is becoming more affordable all the time. The software costs are not decreasing nearly as rapidly. Software will be expensive for the foreseeable future. So the hardware should be selected to minimize the total cost of the system.

REFERENCES

1. Glass, M., "The Meeting of Two Realms: Lessons from the Tevatron Front End", Nucl. Instrum. and Meth. <u>A247</u>, 133-138 (1986).

108

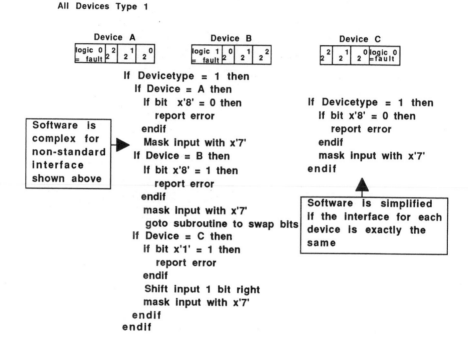

All Devices Type 1

Device A

| logic 0 = fault | 2² | 2¹ | 2⁰ |

Device B

| logic 1 = fault | 2⁰ | 2¹ | 2² |

Device C

| 2² | 2¹ | 2⁰ logic 0 = fault |

```
If Devicetype = 1 then
  If Device = A then
    If bit  x'8' = 0 then
      report error
    endif
    Mask input with x'7'
  If Device = B then
    If bit x'8' = 1 then
      report error
    endif
    mask input with x'7'
    goto subroutine to swap bits
  If Device = C then
    if bit x'1' = 1 then
      report error
    endif
    Shift input 1 bit right
    mask input with x'7'
  endif
endif
```

```
If Devicetype = 1 then
  If bit x'8' = 0 then
    report error
  endif
  mask input with x'7'
endif
```

Software is complex for non-standard interface shown above

Software is simplified if the interface for each device is exactly the same

Fig. 1. An example of the value of a standard device interface.

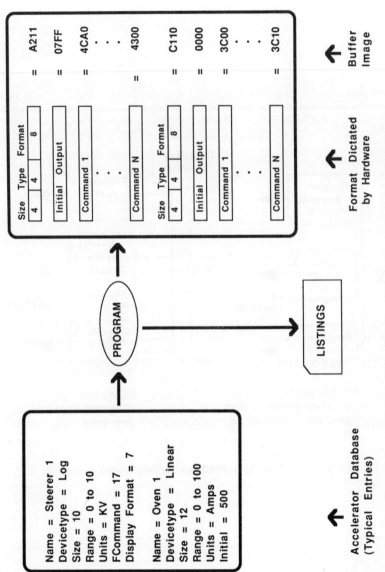

Fig. 2. The value of a database for maintenance of input/output buffers.

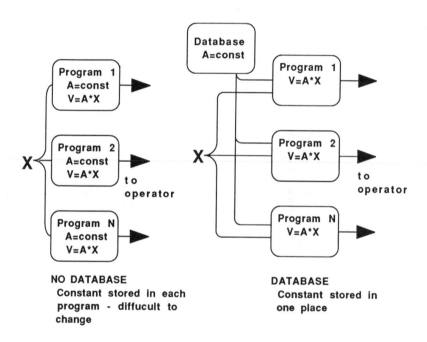

NO DATABASE
Constant stored in each
program - diffucult to
change

DATABASE
Constant stored in
one place

Fig. 3. The value of a database for application programs.

DISCUSSION:

HYDER, Yale: Could you give us some idea of the number of lines of code in different headings? For instance, the number of lines of application code, as in the calculation of foil life times, or the number of lines of code required to drive the console?

JURAS: That's a difficult question to answer. The listing for the operating system is a few inches thick, and if you listed all the application programs it would be about that thick. There's a lot more software for the application programs than for the programs to control the accelerator. Probably half of the programs to control the accelerator are control console interface because that's where all the special devices are. In the accelerator you may have a lot of A/D converters but they're all the same.

Most things require just a linear mx + b conversion, in which you get the x value in and you multiply it and add an offset. A power supply will give you a zero to ten volt signal for zero to ten kilovolts, or zero to a hundred amps, for a meter. When you get to the control console, you have special things: you have the graphics display, you have to format the data for the operator, you have to read the shaft encoders. That software adds up to quite a bit.

McNAUGHT, McMaster: You mentioned bar graphs. Were those analog indicators?

JURAS: Yes.

McNAUGHT: You would not suggest using bar graphs as a replacement for the meters that people like to see, or would you?

JURAS: I don't see why you couldn't use them as a replacement. The problem used to be the resolution, because you had only a few discrete steps. Now you can have a large number of elements in a bar graph, and it responds more quickly than an analog meter because the meter has damping. You might want to put an RC filter on your bar graph to damp it.

McNAUGHT: Are you suggesting that if you were doing this again for Oak Ridge that you might not use meters, you might use bar graphs?

JURAS: We're re-doing our control console right now, and we're thinking about this. I bought a bar graph and connected it to one of the analog meters so the operators could use either one. They're divided about which they would rather use. I think the majority opinion is that they like the meters better.

PARDO, Argonne: At Argonne we've used a four by four touch panel for many years. I still find that that's an excellent way to access calculational programs or major automatic components of a program. For controlling a particular item, your comment about the imbedding and how many pages one has to go through to find the item you want makes a valid point. I think a touch panel with that sort of matrix size is not appropriate.

For beam line controls, we've now gone over to a graphical CRT with a touch-sensitive overlay screen The entire system is divided into fifteen or so pages, and just indentifying the item you want selects it and assigns it to a knob. I'm not claiming our approach is optimum -- I don't think it is yet -- but I think there are at least two complementary systems that work very well.

That system I don't think would work as well as the matrix touch panel for major program location, but for selection of elements on a section of beamlines being tuned I think the graphical identification has worked very well.

Other things I don't like as well, like our up/down buttons. My vote is down on up/down buttons.

CARNES, Kansas State: You mentioned using a database. Are you talking about a commercial database or is that just a parameter table in Fortran?

JURAS: You can buy commercial databases; some people do. We've written our own and it's not a difficult task, and your own database can be more tuned to your accelerator. If you buy a general purpose one, you have to invest a lot of time in understanding how the thing works. If you write your own, you have to invest the time in writing it, but you can tune it to the particular requirements of Tandem accelerators.

We actually have two different databases. One is to generate the CRT pages, so it's in a source format. And the application programs can access these. The database generator program generates the image code that goes in memory that the control computer uses to put the display up for the operator, so the operator can interact with it. But the application programs can also access this database, so that if the operator says, "I want to do something that's on page 27, line 3," the application programs can know about that. They can talk to each other.

ROWTON, Los Alamos: Some of the mouse-driven pull-down menus will allow you to go from level to level within a nest of menus, but they also will allow you to skip all that and call up a particular menu from a keyboard. Is that not an easy compromise for your system?

JURAS: Right, that's a good compromise and something the Vivitron people should be looking at. Another thing is going back one step. Some pull-down menus allow you, after you've got several levels deep, to go back to where you were. It remembers what locale you were working in and gets you back to that locale without having to go back to the top of a menu and go down all the steps each time.

McKAY, Yale: One of the applications that we have considered is access of the database from home for those 4 o'clock in the morning calls.

JURAS: That's one item under control systems that I didn't put down: the disadvantage of computer-based control systems. [Laughter]

McKAY: I was thinking that it might be an advantage if you could call from home and see what the thing is doing rather than asking someone to go and look at such-and-such a meter. Have people provided that?

The next question is: Do you go one step further and send the control to your home computer so you can make a few adjustments in the early morning hours? I'd like to know what people have done and what their reactions have been.

PREBLE, Stony Brook: At Stony Brook that was discussed and I'm glad to say that we stayed away from it in the end. I get enough calls as it is. If I could tune a resonator from home, I would get called ten times a night. [Laughter]

I started working on setting it up so I could warm up the ion source at home while I'm drinking my coffee and again I decided that this was just asking for trouble.

McKAY: Well, what about just being able to look and see what the situation is?

PREBLE: We don't have operators at Stony Brook -- we have under-graduate students and graduate students, we have post-docs, we have any number of people sitting in front of the console who don't really know what it is they're looking at. They're more than happy to call up and say, "Help!" They're not very good at interpreting their surroundings, so they may go into a panic when there's absolutely nothing wrong. I don't want to deal with that at 4 o'clock in the morning. It's scary to give somebody the ability to call you up and say, "Is this right?" You look at it and say, "Yes, that's right, don't worry about it", and go back to sleep.

ROWTON: I'm very lucky that I have a very good staff and I get very few calls in the middle of the night. I would like to be able to look at the parameters, but I'm not sure my wife would put up with having a terminal next to the bed. [Laughter]

GRAY, Kansas State: I think all this discussion points to a general question: Who is in control? Maybe it's part of the evolution of the control process for computers, but when computer control is good, it's very, very good. And when it's bad, it's very, very bad. There seems to be no in-between. There are no options when you are dedicated into a strongly coupled computer control system.

I certainly know what it is to feel frustration when there's not something wrong and you don't know it and you think there is. And I think that really the problem for me has always been in the area of documentation. When all of the facts are packed into one or two people and it's not written down somewhere, it makes it extremely difficult.

It doesn't make any difference then how many computer consoles you have. It doesn't make any difference whether they can call you in the middle of the night, whether you can control it from your bed or not. It just makes a lot of problems. And I think that's an area that we all talk about but, like the weather, we don't do much about it.

JURAS: That's a good point. It's very difficult to keep up the documentation on a system like this.

PARDO: We've had now nearly a decade of computer control experience on the linac with the ability for me to do some things at home, as, inspection of things. If you've got a modem at home you can change a faulty program if you know what's wrong. I've done that on the order of three or four times a year with various things that have gone wrong. That capability is really quite useful if you're twenty miles away. The ability to inspect the condition of the machine has proved useful maybe five times in that nearly ten-year period.

Unless you have invested huge sums of money, you have not really put everything into the system. The way ours has evolved, there are many pieces of the system that aren't totally integrated into the computers. Maybe you'll have everything in. If so, I'd like to see the budget!

But, inevitably, being there gives you so much more information than you can get from home that I found that if there was really a major catastrophe in the hardware, I've never been able to fix it from home. If it's a software problem, then almost always you have the tools at hand to fix it from home or from any other terminal.

MEIGS, Oak Ridge: I just want to say to Tom Gray that documentation does mean a lot. I've been a new member of the staff at Oak Ridge for just about two years, trying to learn the control system, and you do have to have documentation. For the most part they have been very good at Oak Ridge, in software and in hardware.

The other thing that we do for our operators is to try to give them as much information as we can for when something goes wrong, such as a power glitch. They know how to warm-start the computer, and do a cold start, and such things -- we give them as much as we can to fix something, and hope that we don't get as many calls.

ROWTON: There is one real problem that I have with a computer-controlled system that has one screen with a few meters assigned to it. Frequently, I will walk into the control room, have a glance at the console, and by instinct I can walk over and turn one knob to correct a problem. How are you going to do that if you've got a hundred-page menu and you're only looking at one page?

JURAS: Maybe other people ought to say this, but I think our operators are pretty comfortable with the way it is. You can get used to anything.

ROWTON: But you don't have all the information.

JURAS: You don't have the information. That's why things like the alarm panel are very useful. You can set the alarms for thresholds on things like vacuum levels around the stripper in the terminal. It tells you that something is not where you think it is.

ROWTON: Well, maybe, but sometimes it's a very minute change. It isn't that anything is really wrong, it's just not optimized.

JURAS: It would be nice to be able to build a larger console. I think that's a trade-off that you get into with a computer-based control system. It buys you lots of things but then you have to think very hard about how to make the new console. You could, as they do in airplane cockpits, simulate the old controls, just make them computer-driven. But that's pretty expensive.

ROWTON: And it defeats some of the reasons for doing it. But I certainly like having all the information that I can see at a glance.

JURAS: More information. Well, it's always possible to come up with several CRT displays.

STORM, Washington: It's important to keep in mind the size of what you're dealing with. If you're dealing with an FN Tandem with an analyzing magnet and a couple of quadrupoles, you ought to have it all on meters so you can look at the whole thing at once.

If you have a system that needs a hundred pages, that means you probably have a hundred or two hundred devices and you're not going to have them on meters that you can keep track of with your eye anyway. I think it's important to understand when the system has gotten beyond something that you can absorb at a glance and then to organize the menu system, if that's what you have, so you can look at the sub-systems in a way that you can look through a sub-system right away and understand whether it's working reasonably.

LINDGREN, Brookhaven: On John McKay's question, about getting back into the system from home, the Brookhaven experience has been that it's the computer system that fails. That's where we need the help, not with the HITL lines or the Tandem facility.

The other thing is the NMR's. In modern NMR systems you no longer need the oscilloscope panel to see the trace. The Bruker units we have just lock on automatically and read back the gauss settings.

McKAY: I think we have come to the conclusion that we want a number of displays for these functions. The alarm function is one that you want either on a separate display or set up to interrupt the display you're looking at. One of the things we've talked about somewhat casually at this point is a separate screen that's just sort of meandering through the database and displaying these things, so when the operator is there monitoring he can start flipping through it. That's not quite the same thing as being able to walk in and spot something wrong anywhere on the console, but it does allow that casual checking of what's happening as you go.

SOUTHON, Simon Fraser: Something Bob [Lindgren] said brought up another point that I think is important. Maybe some of the people that have control systems can tell us what they've done to make sure the system goes down gracefully when the computer bombs and where they have hardware backup and how much.

JURAS: There are things that can be done. When our computer goes down, often the beam is still on target. It can go down with the beam still going through.

Bringing up the computer -- there are nuances of control. For example, you can have a power supply that stays on normally, but you have to give it a momentary control signal to turn it on or a momentary control signal to turn it off. We try to do that now.

Another way is to have a register bit control the supply, so that if the bit equals one, the power supply is on, and if the bit is zero, then the power supply is off. We find that the former type of control is better, though it requires a little more wiring and attention. We also have buttons in our system. There are buttons to turn the chains off or to turn the rotating shafts off if the computer dies and you have no control of anything. We can shut it down gracefully.

One problem that we have is not so much computer related. There is a leak valve at the terminal for gas stripping. And if we lose the rotating shaft for some reason, then the tubes get pressurized. I don't know of any good way to get around that.

HYDER: I suggest quite a good way to get around that is to use the HVEC thermo-mechanical leak which closes when the power disapears.

GRAY: Certainly we might borrow a page from the airlines, or the aircraft industry, where there are a pilot and a co-pilot for landings and take-offs. Once in the air they can go autopilot all they want. But when they are close to disaster they usually have an interactive control. The computer doesn't run the whole show.

JURAS: I believe the computer in the modern airplanes does do the take-offs and landings.

GRAY: But there is still a steering wheel and a switch that says on and off.

JURAS: There is still a steering wheel, right. I was talking to a pilot on a flight recently who was training on a 767. He said that the pilot is not allowed to land the airplane because they feel the computer is more predictable and reliable.

GRAY: I hope they don't have a disk crash 15 feet off the ground! [Laughter]

JURAS: Right. The pilot's there for emergencies. But he said his feeling about that is that he may get rusty and complacent, and he doesn't know if he'd be sharp enough to take over. He doesn't have the continual interaction with control of the airplane (or the accelerator in our case).

SOUTHON: Ray [Juras], do your radiation monitors go through the computer or are they separate?

JURAS: They are separate. A lot of that had to do with how the system was built. NEC built the control system while we were doing the rest of the building.

STORM: This may or may not be applicable to people with Tandems, but in our linac we have ten satellite computers that are either DEC Falcons or PDP 11-23 pluses. They are controlled by a main computer. They stand alone. So if the main computer dies or gets turned off or gets unplugged, the small computers keep on doing their jobs. They're told what values things should be at and they keep them there.

You can go locally and turn devices off. The small computers are connected in such a way that if they die the systems will die in a reasonable manner.

For example, the vacuum system has its own computer. The interface for the vacuum system monitors whether the computer is talking to it or not. If the computer stops talking to the interface, the vacuum system dies gracefully, the pumps turn off and so forth. The other systems don't die, they just keep going.

These little computers are very cheap and very convenient. You can have lots of them and have one for a given system and then treat it the way you want to.

JURAS: Our computer system has evolved in that way also. I didn't have time to talk about the trade-off of a big host computer versus a distributed control system, but we're doing that too with small computers right at the front end. Even though they're in assembly language, the code is usually so small it can still be a very reasonable thing to do.

STORM: For the DEC Falcons we write the code in Micropower Pascal and we have about one or two percent of it in PDP assembly language.

McKAY: I understand from Neil Burn that the magnet set-up at Chalk River now is done completely by the computers. Their original intention had been to have the computer do the first approximation and have the operators do the fine tuning, but they found that the computers, carefully programmed, can do a better job than the operators can. Perhaps it is time to yield gracefully to the fact that the computer can do a good job in these things.

HEIKKINEN, Lawrence Livermore: You implied you are going through an upgrade modification of your control console so you're not really starting from scratch. How much manpower is invested, in particular with respect to software?

JURAS: There's quite a bit of software to be re-done, but there are only Martha [Meigs] and I to do the software, with a few college students, or whoever we can get. It will take a couple of years, or a year, along with other jobs that we have to do. But we're only talking about a small amount of the console, not the whole thing. We're not going to throw away the old one. We'll keep a lot of software.

DARLING, Rutgers: If you're going to set your machine up so you can access it from home, how do you keep the hackers out? And, what do you find that you need in the way of accuracy on your most accurate system?

JURAS: Most of our D/A converters, because of what we could buy ten years ago, are ten bit D/A converters, for steerers. There are a few functions that have 12 bit D/A converters, like the terminal reference voltage. The magnets have 20 bit D/A converters, but actually 19 bits are used. We find that that resolution is probably necessary on our mass analyzing magnet. We have a long way to go up the column so there's a big moment arm on it.

DARLING: How do you implement that level?

JURAS: That level is done with mercury-wetted relays with a sort of digital version of a Kelvin-Varley voltage divider, with 20 stages. We have a box made by North Hills Electronics.* NEC bought it for the control system, and it works very well. When you update it, you find that it has a glitch every time you write to it, but the relays are latching relays, so once you write to it it's very stable. We place it in the power supply so that the reference voltage is divided there and it doesn't have to come outside to a CAMAC crate or someplace where it can pick up noise and ground loops.

TRIMBLE, Florida State: The way around the hacker problem is quite simple. You make your modems originate only so they will not accept any incoming calls.

JURAS: Right. You have to call the regular phone and ask an operator to call you at home and start it all up.

BERNERS [Chair]: It appears that there are no more questions on that topic. The rest of this session is for discussion of laboratory reports. I'll ask Bob Lindgren to begin that. He has some pictures that he would like to show.

LINDGREN: In my report I talked about two accelerator tubes we pulled out of MP7. It has extended accelerator tubes, and the whole set was installed in about 1981. Tubes 7 and 8 were giving us trouble. In individual conditioning, we couldn't get as high in voltage with those as with the others, so we changed them.

Tubes 7 and 8 had numerous glasses damaged. What concerns me most is the damage on and near the spark gap electrodes. All of the glass damage is on the top half of the tube. The top 135 degrees of the tube has a lot of damage on the outside, all the way down the tube, and each of these tubes has four or five glasses that are damaged inside the glass. Each place with inside damage corresponds with damage on the outside of the tube, leading me to believe that the outside damage came first, and that repeated sparking along those tracks caused the damage inside the glass.

---

* North Hills Electronics, 1 Alexander Place, Glen Cove, NY 11542 USA. (516) 671-5700.

There was a lot of debris on top of the tube. I can only believe that this debris is dust from pulleys that have disintegrated, perhaps gathered in an oil film from the oilers on the sheave assemblies on the Pelletron. There is no damage at all on the bottom of the tube, and that's the first time I've seen tube damage like that.

On those tubes that we took out, about 50 percent of the spark electrodes were loose in the holes in the tube electrodes. On inspecting the spark gaps I saw another bit of damage on a good many glasses. It is a little circular bit of disintegration of the glass, underneath the spark gaps. That did not occur underneath all spark gaps. It did not occur only on the loose spark gap electrodes. I think this damage occurred because of the proximity of the glass to the spark gap itself. I think they should be further apart.

I have also seen burn marks around the spark gap electrodes, where a ring is pressed onto the electrode from the opposite side. These marks occurred on perhaps 25 percent of the electrodes on the tube. The spark went from the spark gap electrode across the ring, and then to the tube electrode itself. To me, that indicates a fairly high inductive connection between the spark gap electrodes and the ring. We have had some of the spark gaps have the holes eroded out completely so that the spark gap is wiggly.

I think that there should be a better way of anchoring spark gaps into the tube electrodes. They should at least be a press fit into the electrode. HVEC drops them into the hole and then presses the compression ring to make a tight fit up against the electrode, but that doesn't make them have a good joint.

We have made some different spark gaps that make a better connection, like the type used on the Dowlish tubes. We have put spark gaps like this on the HVEC tubes, because in order to hang our tube resistors onto the tube we have removed some of their spark gaps and just left the hole there, and we anchor our resistor mounts onto the holes in the tubes.

So we do have a two rows of these down the tubes, and the advantage of these is that they can be mounted further from the glass. They will not give you the damage on the glass itself.

BERNERS: Bob, did you have any damage that penetrated through to the vacuum?

LINDGREN: We had no vacuum leaks.

ROWTON: Some years ago I noticed that some of the spark gaps on our FN tubes were getting loose and it seemed that the interference fit on that pressed-on ring was not sufficient. I made up a simple tool with a pair of vise-grip pliers and re-pressed them, and I think they have maintained contact pretty well. I don't see the marks like you showed where it's burned underneath the spark gap itself.

LINDGREN: This is the first time I've observed that also. And we have also made a tool so that we can take these spark gaps off, put them back on and then squeeze the ring back up, but that doesn't change the fact that the spark gap itself is a slip fit into the hole.

ROWTON: But one side of your spark gap is in good, firm contact with the electrode.

LINDGREN: But there is no continuing pressure to hold it there. When that ring relaxes, it's no longer tight.

ROWTON: But if you made tighter rings and pressed them on --

LINDGREN: Ah, I spoke to Charlie Goldie about that and he assured me, oh, yes, when we rebuild a tube, we always replace the rings. And so I can only conclude that the rings are not tight enough on the electrodes, because they are coming loose.

When we were at Yale we felt the spark gaps on the column and on the tubes, and we felt a few loose ones there. So, they have not done a perfect job yet.

ADAMS, Pennsylvania: That last vu-graph you had with the spark going out radially from the spark gaps. They seem to be on one side of the spark gap. Now, is that the cathode in respect to the previous electrode or the anode?

LINDGREN: I've observed them on both sides. At first I thought they were only on one side and I can't say which is the cathode and which is the anode. But then that would have been reversed if we ever ran the machine with a negative terminal, wouldn't it? So, I can't answer that.

HYDER: Another question on the same topic. What was the closest column component in the direction going out, in the direction of those radial spark marks?

LINDGREN: The Pelletron chain.

HYDER: So, presumably this means that during some surge condition there is enough of a transverse field to the column to cause sparking from those gaps out to the chain.

LINDGREN: That's the only thing I can see that's in that direction, Dick, from the marks on the side of the spark gaps, but it seems so absurd because the distance has got to be nine inches. And I just can't imagine that the sparks are going that far in the insulating gas to the tube electrode.

HYDER: Well, there are several million volts available.

LINDGREN: Yes, fifteen.

STARK, McMaster: Could you tell us a little more about the foils and the annealing process?

LINDGREN: In my lab report I mention the pulsed beams we are injecting into AGS. We never noticed this problem with foils when we were running oxygen and silicon, but we were doing some runs with gold and we were injecting 80 microamp pulses of gold. The repetition rate on these is about two per second. The duration is perhaps 300 microseconds.

When we first lifted a cup out and let the beam go into the terminal, we thought we were in the position of a good foil, but we got nothing at the high energy end. So, we put another foil in and still nothing. Finally someone had the idea that maybe we were breaking the foils as soon as we put the beam in. So, the intensity of the beam was backed off to about five microamperes and then we could see it on the H. E. cup.

We found that if we gradually increased the intensity of the beam up to its full value, taking anywhere from 30 seconds to a minute to get to the full value, the foil would not disappear. In fact, it had a lifetime of up to six hours. These were slackened, two microgram/cm^2 foils.

I can only assume that it's annealing, and I use that term because we're guessing that the material is being annealed by the beam before it has a chance to break.

CHAPMAN: Some years ago, Yntema at Argonne had a scheme actually for heating the foils and claimed that he got an extended lifetime, by just such an annealing effect.

LINDGREN: Yes, but it wasn't reproducible, and I think this phenomenon that we mentioned is reproducible and so it's a different type of heating.

ROWTON: If I remember right, carbon is one of the strange materials that gets stronger as you heat it up, within certain limits of course.

LINDGREN: I think most of our foils are standard evaporated carbon foils, and we haven't found any great difference in lifetimes with cracked ethylene foils or with evaporated carbon foils. But they do want to be slackened. We find that's a big help.

ROWTON: We're still slackening our foils with our xenon flash gun. That's very easy and convenient and I definitely see a difference in lifetime that way.

BERNERS: Can you describe how the xenon flash gun slackens a foil?

ROWTON: No. [Laughter] All I know is that it does. There were a fair number of experiments done at Los Alamos many years ago before I joined the group. They indicated from several of their tests that the lattice becomes more diamond-like rather than amorphous. You can hear the foil make a sound when you flash it. At first it's nice and tight and shiny, but after flashing it gets loose in the frame and frosted in appearance.

BERNERS: What is the power output of the flash gun?

ROWTON: It's a standard photographer's flash, not a weak one, but a reasonably powerful one. We place it about two inches from the foil after we've already mounted it into our chain, flash it, rotate to the next position and put in the next foil. I've reported this to SNEAP several times.

BERNERS: Who makes your foils?

ROWTON: Arizona.* The last batch we flashed were very poor. We had great difficulty getting enough foils from two boxes of slides to fill our chain. Our chain holds a hundred foils. So, the release agent is poor at this point.

SOUTHON: I have a comment in the form of a story. There does seem to be some kind of thermal shock effect associated with first putting the beam through. I do mass spectrometry at McMaster and typically we run carbon beams of several hundred nanoamps of ^{13}C on one sample for perhaps one or two minutes. Then there would be a period of 10 or 15 seconds while we change to another sample. We pop the cup open and let the beam through and a surprisingly high percentage of the foils break as soon as we let the beam through. The foil has had time to cool down, perhaps, and then the beam heats it and shocks it and it goes.

The other thing I'd like to mention is some rather strange behavior that we've observed in our foils. These are slackened, cracked ethylene foils, typically five or six micrograms/cm^2. We have found that over a period of several hours of running, these foils will thicken up. We see a higher percentage of the beam going on to the image slits and we start to lose transmission down the beamline.

---

* The Arizona Carbon Foil Company, Inc., 4152 East Sixth Street, Tucson, AZ 85711 USA. (602) 624-1881.

So, we switch to another foil because at that point our carbon data will start getting unreliable. We find that if we come back to that foil two or three weeks later, it's gotten thin again.

The other strange piece of information is that several months ago an experimenter, who shall remain nameless (but it wasn't me), was running one of these carbon dating experiments and found that the foils that had been put in were not Stark Specials, but they were made by a trainee and they were extremely thick. This guy got thoroughly annoyed at this and decided that Operations were going to change all the foils and they were going to change them because he was going to break them all. So, he cranked up the source and changed the injection system and put 25 mikes of carbon on these things.

He couldn't get any of them to break, so he gave up after five minutes or so on each foil. Then, when he came back to them again, they had gotten thinner.

So, there's my story. If anyone has any explanations on it, I'd be very interested to hear them.

BERNERS: John, who is the maker of those foils?

SOUTHON: Jim Stark. He made the good ones, not the bad ones. I think we know that it's not just a straight cracking effect. If hydro-carbon vapors or whatever actually cracked out on the foil, I can't see that they'd get pumped away over time.

HYDER: If foils get thicker this is beause of deformation or accretion, presumably. First of all, deformation of the original foil so that something that's originally slack becomes irregular and taut, thicker in places. This is clearly limited to the extent to which you can redistribute the material in the foil. The second reason for the foils getting thicker is that material has accreted onto the foil, and that must be dependent on the local vacuum and the time.

Foils get thinner because material is being sputtered from the foil and that's basically dose dependent, on the number of particles striking the foil. So if you put on a certain dose in a very short time, you lose a certain amount of material because of the sputtering, but the accretion during that period is small because the time is short. If you put the same dose on much more slowly, the sputtering loss will be the same, but the accretion will be much greater. One assumes here that accretion only occurs when the beam is on the foil. In other words, the beam does something to the foil. It charges it or excites the material in the foil so that accretion takes place when the beam is on the foil but not when the foil is inactive. So, I think that can explain part of what John described. What I didn't quite understand about Bob's remarks though was this: that any heating effects are presumably short in time compared to the intervals between successive pulses, which is usually half a second. So, was your long life a result of some permanent change in the foil that arose during the initial bombardment of low intensity?

LINDGREN: We don't know what it was. I'm presenting it to you. I was hoping to find an answer here.

BERNERS: Perhaps someone remembers the details of this better than I do. At some past SNEAP a delegate to the meeting from a European laboratory described a lot of work that had been done on the thickening of foils and said that there was another mechanism, which was just a migration of carbon atoms into dislocations in the foil that were produced by beam bombardment. Isn't that a third mechanism, in addition to the two that you were talking about? And what does this have to say about what John was describing?

HYDER: Well, this is the rearrangement of the material in the foil. Essentially you're transferring atoms from one place in the foil to another. And I think the evidence is that if this happens the foil becomes less uniform. But the extent to which this can affect the foil thickness must be limited by the total amount of material available there to be redistributed.

SOUTHON: I'm sure this happens to some extent, and I would have thought that effect would be frozen in once the foil has been hit. So, that may happen but I don't think that has anything to do with the foil getting thinner again.

ROWTON: I believe that was a Daresbury report. And this was one of the mechanisms they were hypothesizing as the root cause of foil breakage. You could migrate enough material up from the perimeter to make it so thin that it would rupture.

GRAY: Several years ago we studied the ablation of solid materials from the surface of solids with heavy ion beams. We saw in those works (which are published in the Proceedings of the Denton Accelerator Conference of a few years ago, and Barney Doyle I think is the principal author of that paper) that the material comes off at 90 degrees with respect to the beam. We had a 45 degree tilt on the surface and we were looking for the slow velocity components of ablated material. These materials come off the surface with an energy of the order of electron volts, so they're very distant collisions. We looked at both the front and the back sides and saw this kind of process going on. It is sputtering, and the material is moving basically at right angles. If you have a tilted foil you would accentuate the loss of material over a flat foil arrangement.

In a wrinkled foil, I can see where you'd be catching some of the material that you're throwing out. It's not energetic at all. It's not a hard impact collision.

SESSION VI

DISCUSSION OF LAB REPORTS

Mr. McKAY (Yale) Chairman: The next bit of this Session is to carry on discussion of the lab reports. I have a few things that I've looked at, but I would rather ask for questions from the audience.

You've all had a chance to collect the lab reports and I know that most people spent the evening last night going over these in detail, highlighting points of question and so on —

(Laughter)

and it would be presumptuous of me to interfere with that work. So, I think that perhaps I'll just throw it open to the floor and see what happens at this point.

Mr. BLANCHARD (McMaster): I noticed VIVIRAD resistors. I'm not familiar with them.

Mr. McKAY: The question was, information about VIVIRAD resistors. VIVIRAD resistors are the ones that Michel Letournel has developed. Before I let him have the floor, I noticed a couple things in the lab reports relating to this. You may recall that last year Neil Burn from Chalk River reported some serious problems with them. I would point out that in this year's report from Chalk River the problem has been corrected and they've been quite satisfactory. We have them in the ESTU and so far they have gone to around 22 million.

Beyond that perhaps, Michel, you would like to make some comments about developments with those resistors and a brief description of them.

Mr. LETOURNEL (Strasbourg): Yes. We had some difficulty with a type of resistor sent to Chalk River and to another lab also. It was just a question of varnish, which a subcontractor had developed in a wrong way.

It took us quite a long time to find exactly what the trouble was. During that time we worked in different directions in order to improve those because we were thinking that was not possible. In fact, in addition to the correction of the varnish we came up with a product which was better, finally. The protective end-caps are much smaller and they are in stainless steel, completely stainless steel, and they are better electrostatic design.

I would like to add that essentially the protection, the frame which is around the resistor is as essential as the resistors themselves. Above 15 MV, we have deterioration in many components of a big accelerator. Working with different types of frame and protection, we came up with two kinds of frame, and now on the Vivitron we are going to put resistors on each side of the tube, not anymore above or below. The frame will be much simpler. I think that Padova already has them for some part of their tubes.

Mr. BLANCHARD: Stony Brook does digital megging of the resistors. Could they point out how they do this? Is it a commercial item or do you supply a high voltage and put it to a volt dropping resistor on a digital meter?

Mr. WESTERFELDT (T.U.N.L.): I believe that John Noe included that in proceedings several years ago. It uses a small plotted high voltage power supply, I think 5,000 volts.

Mr. LINDGREN: (Brookhaven): I've got the information from John Noe. He makes two tests. One is with a three kilovolt power supply with the current calibrated in megohms.

But his ultimate test is to measure the resistance at nine volts. And for this test he uses a Fluke true RMS voltmeter that has a siemens scale on it. If you try to measure 800 megohms with an ohm meter, it takes perhaps ten minutes for the digits to go up to 800 megohms. Whereas, you can read the inverted ohms, siemens as they call it now. The meter is calibrated in siemens and it gives you numbers in those units which can be converted to one over ohms or to ohms or megohms.

That meter is a good meter and has about four digits so you can read the resistance very carefully.

We have tried using that and it shows up resistances that are open at nine volts. They would not be open at five kilovolts, but John's claim is that that is a much truer picture. If you read 765 megohms with that siemens scale, that's what you've got.

Mr. McKAY: I'd like to comment that measuring resistances and assuming that Ohm's law works is probably a mistake. We're measuring our resistances with up to five kilovolts (depending on the humidity) using a Criterion Research Instruments, five megohm portable megger.

But if you test resistors for resistance against voltage, you get some most distressing curves. Perhaps you should pick a voltage and only test at that voltage and not look at the variation.

Mr. JANZEN (Queens): I've just gone through this exercise with our resistors. Those are the old high voltage blue epoxy resistors. Their nominal values are 240 megohms for the low ones and 400 megohms for the high ones. I've done both of these tests, one at 20,000 volts on the cross resistors in air and one at nine, and measuring the current at both of these voltages and then calculating what the resistance was. I was extremely surprised to find out how consistent the values were at 20,000 and at 9 volts. And not only that, they were always the values specified. In other words, 400 megs and 240 megs. So, your comment about this change in resistance with voltage may be valid, but certainly for these blue resistors it's within a few percent.

Mr. STARK (McMaster): We have Caddock resistors in our FN accelerator, and after John Noe's comments about the problems with microbreaks, we watched for them with a meter and I've replaced maybe seven or eight resistors in two years.

The biggest question is that I haven't seen any problems in the operation of the machine. I'm sure not too many people run at nine volts on their column resistors, so, is there a problem with microbreaks? Does it cause mechanical failure?

Mr. PREBLE (Stony Brook): Okay. You look at nine volts to identify these microbreaks and the idea is that the microbreak is a predecessor to worse problems down the road where you will develop an open. So, that these nine volt tests are giving you signs of things to come or can give you signs of problems that will be produced by these microbreaks.

Mr. LINDGREN: I have often felt that that's not the real world. The real world is putting 40 kilovolts across them. I'll go along with John, it's a very nice way of measuring the resistors, but I believe that we operate at above 25 kilovolts and it's going to break down in the opens.

However, we're listening to them. And I agree with Henry Jansen that the resistor checks give you a very good idea of the resistance. We have at times found when we go in the machines and just flip across the sections, we may get a high current reading, which when we remove the resistor from the tube we find that it reads alright. This we put up to the fact that there is perhaps moisture in the dust on the tube which is bypassing the resistor. But in general, you can get around that by just removing it from in place and now you're not reading moisture. At least there is no seepage across the body of the resistors.

Mr. STURBOIS (Ohio University): I think that stability is tremendously important if you're going to run it at high voltage or high voltage gradients. Any time you develop a resistance that you rely on bridging with voltage, you create a potential spark gap situation. And when you create that, it may look stable on a DC meter but in the RF frequencies, you get a lot of instability. I think those instabilities really hurt a machine when you run it at high voltage.

Mr. PREBLE: We rely very heavily on the stability of the resistor. John has worked very hard with various resistor designs, mounting designs, and we have now adopted the Brookhaven resistor design and are building these. They seem to provide tremendous stability, and so our machines are working well.

Mr. BERNERS (Notre Dame): A question for Joe. Have you modified the Brookhaven design in any way or are you using it as they last reported it?

Mr. PREBLE: Very slight modification. In our beta patching process, there was a little bit of difficulty getting a nice alignment at both ends of the tube. So, there is still a little bit of thought going on in possibly modifying the attachment point to the middle of the tube. And the coil at the end around the lucite has been modified very slightly. But it's essentially the same design.

Mr. LARSON: I would reinforce the comment about stability. Especially with inclined field tubes you need very stable operation from your resistors.

Secondly, it's implicit in some of the discussion that has gone on but no one said it directly that when the resistors are properly protected and survive well in the environment, the measurements need to be even more precise than in the past. Measurements to within one percent accuracy or so I think are desirable now and you can't get that out of some of the old megger instruments. You need digital metering.

Mr. McKAY: Are there other comments or topics that people might want to bring up from what they've read in lab reports? Yes.

Mr. CARR (Caltech): We have a comparatively new Pelletron and we still have the original corona points down to the column. I see in various lab reports that people have changed them from time to time. That's a fair amount of work. I'd like to know how do you know that it's time to do that?

Mr. ZIEGLER (Oak Ridge): When the current is drawn through the corona point, drops significantly, I think it's time to change it. We normally change them around once a year, but that's four to five thousand hours.

Mr. STIER (National Electrostatics): While I was at Sao Paulo on the 8UD we were able to run the corona points until they were pulling about 30 percent or 50 percent less than their initial value. And at that point, we'll change them out. And they could run a year or 5,000 hours to 10,000 hours. Now, I'm speaking for the laboratory at Sao Paulo. I'm not speaking for National Electrostatic Corporation.

Mr. TESMER (Los Alamos): I suggest changing to resistors.

(Laughter)

Mr. McKAY: There is a topic that I would like to bring up. I've heard some comments about recent problems with various turbo pumps and I'm interested in this as we've purchased some Leybold equipment and we are talking about getting some Balzers equipment. I wonder if some people would like to mention their recent experiences with these two manufacturers.

Mr. GRAY (Kansas State): Thursday I had planned to talk a little about our experience with Leybold-Heraeus greased bearing pumps, TMP 360. It has been disastrous. Chrysostomos Hadjistamoulou has put together a rather detailed history, hours of failure point, et cetera. We have greater than 50 percent failure rate.

Mr. STARK: I just got a call last week. Leybold are pulling the grease stick off of the 360 and going to oil. They won't be producing the grease stick because of bad lifetime.

Mr. ZINKANN (Argonne): We have the Sargent-Welch six-inch pumps. I think 300 liter. We have four of them and I don't think that we have all four of them working at one time. They're very bad. If you don't want them to work hard, if they're looking at 10^{-7} beamline all the time, it's okay. But if you try to pull the beamline down a few times, that's it. It's gone. We've had very poor luck in running the small pumps.

Mr. ADAMS (Pennsylvania): Last year's SNEAP meeting, I made a comment that we'd been preparing to put a terminal turbo pump in. Our situation is a little different. It's not like some of the SF_6 machines which run at about 100, 110 psi. Our problem is that we have to run at 200 psi.

What happened the first time it went in, we pressurized up to about 140, 145 psi, then there was a beautiful step function on the vacuum gauge and all hell broke loose. We had to immediately depressurize. When we got down to about 135, the leak closed up to some degree. But we had to pump out the tank and pull the pump out.

We did a few things which we thought solved the problem and we put it back in again. The same thing, step function at about 145 psi. Well, you just can't make things work running in and out of an accelerator like that.

We had to make a pressure vessel in which we could stick this 150-liter turbo pump, backfill it with helium up to working pressure and go step by step. Sure enough, at 145 psi we had a step function again.

We took it apart, changed some O-rings in the body where the grease fittings are. We put it back in again, things improved to some degree.

We had to remove the thing, pull the top of the pump off and reoperate the O-ring groove around the pump body and go from a millimeter cross-section over to a standard 1/8" diameter O-ring that we use here, the .130 diameter. Also on the bottom of the pump, there is another flange which they get the bottom bearing in. We changed from a millimeter there and went to a 1/8th cross-section, put it back together again and that seemed to solve the problem. We could get to 200 psi.

Everything was looking real great. So we put it into the terminal, and the bearings went. So, we had to pull the pump again. Our philosophy has been this: The one pump was sent back to be rebuilt and we bought another TMP 150 pump. If it's going to have 500 hours lifetime in the terminal before the bearings go, that's what we're going to have to live with. We also have a third pump in which we are just testing the bearings that come directly from the factory package. The bearings that Leybold-Heraeus have modified cost about $400 a pair. You can buy a bearing, I think Jim Stark can bear this out—just put them in from stock and run them. I might also mention a 360 pump that had 500 hours on it and failed with bearing problems. We had to send them back.

On the terminal operation, we're just going to have to live with it, this Leybold-Heraeus pump, with its grease bearings because you can't put an oil lubricated bearings into the terminal. That's it.

Mr. STORM (Washington): And I have a very simple question which is: When you put your own bearings in, how long did they last?

Mr. ADAMS: They're still being life-tested. They've been on for a week or so at this present time. So it's not conclusive.

Mr. STIER: Have you been cooling those bearings at all? Cooling that pump? We've put a small fan on our units.

Mr. ADAMS: We were worried about putting a regular fan on there because of the drag. Fortunately, the water heater that drives this pump is practically facing the back end of the turbo pump. So, we took the fan off the big CTI compressor, and mounted that on the alternator. That throws up a lot of air at 3500 rpm, and that's what we're using to cool the face of the turbo. We're leaving the grillwork on there, the radiator that's attached to the bottom of the pump.

Mr. HEIKKINEN (Livermore): Over the past nine years, we've had about 15 of the Sargent-Welch 1500 liter per second verticals, six of the big horizontal Sargent-Welch nominally thousand liter per second pumps, and more recently 12 of the 2,000 liter per second Balzers.

The Sargent-Welch verticals, we converted at least eight or ten of them to junk. We converted all the Sargent-Welch horizontal to aluminum shavings. For the past five years we've had Balzers. We had one fail and which it did some rotor damage and that was due to the Balzers' people demonstrating how to rotate the end belt. They did it incorrectly. Just last week we had a bearing seizure on one pump which was corrected within a half hour by one of our mechanical technicians. We do however change bearings every six months. It's a half hour job.

Mr. HADJISTAMOULOU (Kansas State): I would like to comment on the business with the Leybold-Heraeus turbo pumps. To my experience, it doesn't make any difference how you cool those pumps. If they're going to fail, they're going to fail. Since the last year we have purchased about 14 of them and we have five of them on the beamlines. The place we have them is not that cool, and we have air fans on them.

We have five of them. One has 14,000 hours, some have about 7,000 hours, others have 4,000 hours. Tom is going to give the statistics Thursday. And they work fine. I had few of them in the first basement in our department and I have them in the room. I cool them with water and it seems to me they are dying like flies, one after the other.

What I have noticed since I started working with those pumps is that the serial number, looking from the last digit, the fourth one is a letter. All the pumps that we have failed so far have the letter "O". The others have "R", "E" and stuff like that.

So, I called Leybold-Heraeus and they said to me that in the factory where they make those bearings, they got let's say 5,000 of them defective somehow. And they just put them in the box and sent them out. Now they're trying to find out where these pumps are I guess.

(Laughter)

Mr. HADJISTAMOULOU: Now, they gave us two years' warranty on those pumps. Each time we sent the pump back, either they try to repair them or they send us new ones.

Mr. McKAY: This is a very encouraging set of stories.

Mr. PREBLE: We've been dealing with Balzers pumps on our linac and we have a 170-liter pump there. We're changing a lot of bearings, and we found oil in the cryostat in one case. Luckily we have a cup sitting underneath to catch the turbo blades when they fall in.

Balzers was kind enough after repairing some pumps for us, to return them with the foreline inverted so that there was a possibility anyway of backstreaming of the foreline into the bearing oil. It's not been a very pleasant experience. We're in the process of removing them off our linac.

Mr. WEIL (Kentucky): We have several small 50-liter Balzers pumps and we have had them just a short period of time. We were having a problem with the power supplies kicking out. They're on beamline. We don't monitor them.

We've had to go to putting a TV camera on to watch the pilot light to see if they're running or not. I wonder if anybody else has had problems like that. They start up right away again, but it's inconvenient to have your back to the pump and have it quit out.

Mr. McKAY: It sounds as if the interconnecting cables between power supplies and the turbines themselves have been very reliable but the rest of the system needs work.

(Laughter)

Mr. WELLS (Kansas State): Does anybody have turbo pumps that work reliably?

Mr. ZINKANN: We have Sargent-Welch 1500-liter pumps on our cryostats. And the pumps themselves, after several years of working with pumps and fixing pumps, the pumps have been doing quite well. We've had no problem. But the power supplies I think were designed by high school students. Those things go out quicker than we can get up with them. The but pumps seems to be fixed.

Mr. STORM: I'd like to add to the remarks about the Leybold-Heraeus pumps. We have about ten of the Leybold-Heraeus TMP 150's and 350's and we have found that the ones that were bought at first have worked very well and still work very well. This is consistent with the remark from the gentleman over there. The new ones fail. The story that we got from the manufacturer was that the company that makes the bearings had re-tooled and was having trouble now with the tooling making bearings that will work at all.

This is consistent with the fact that apparently they're no longer making grease bearing pumps and are switching to the oil. They've been very cooperative as well as they can be in terms of fixing the newer pumps that are breaking for us. It sounds like it's not going to be a lot of fun.

Mr. McKAY: Thank you very much for your contributions to the discussion.

SESSION VII

GLUE-BOND FAILURE IN DOWLISH
FN TANDEM BEAM TUBES

R. Bundy, R. Lefferts, J. Preble,
C. Purzynski, J.W. Noe, and H. Uto

Physics Department, SUNY at Stony Brook
Stony Brook, New York 11794-3800

SUMMARY

All four Dowlish beam tubes in the Stony Brook FN
tandem accelerator failed in May 1987 when numerous
PVA bonds separated during an extended maintenance
period. This review of the incident will emphasize:
1) follow-up preventative measures and 2) problems
encountered during installation of the rebuilt tubes.

INTRODUCTION

In late May of this year we experienced a dramatic
failure of the PVA glue bonds in our Dowlish spiral inclined-field accelerator tubes. The problem was discovered
when the tubes could not be pumped down after having been
open to a dry nitrogen atmosphere for about 10 days to
complete some gate valve repairs. During part of this period the ambient humidity in the lab was close to saturation
as full summer air-conditioning was not yet available.

Upon inspection of the tubes we found a broken glass in
tube #1 near plane 38 (counting from the beam-in end).
After we removed tube #1 and still were not able to pump
down, we found that tubes 2 and 3 also each had one broken
glass in nearly the same location. Figure shows the typical appearance of a failed joint; the glass has cleanly
separated from the metal about half way around. Tube #4 was
removed intact, but like the other three tubes showed evidence of the start of extensive separation of the glue
bonds over nearly the entire bottom edge of the tube.

This failure mode is similar to that reported at previous SNEAP meetings by Chalk River, Notre Dame and Florida State, except that for these labs the tube age at failure was invariably well over 10 years (rather than 7) and HVEC tubes were involved. As at Stony Brook the earlier failures seem to have occured after tubes were opened for several days or weeks following a long period of operation.

HISTORY

The Dowlish[1] tubes were installed in 1980. As of May 1987 they had operated for about 22,000 hours at voltages up to 9.3 MV. The electrodes are titanium with a 10° inclination angle except for the initial section of tube #1, which is magnetically suppressed with small internal magnets. The overall ion-optic design was done by Hyder[2].

Figure 1.   End view of one of the three failed tubes.

When the Stony Brook tubes were ordered there were only a few Dowlish tandem installations, at Pittsburgh, Los Alamos, and Rochester. Subsequently, the FN tandems at McMaster, Florida State and Pennsylvania have converted to the Dowlish spiral-i.f. design with reportedly good success. Tube performance at Stony Brook prior to the failure was also quite satisfactory, especially in our application where the tandem serves primarily to inject heavy ions into a booster linac. The improved vacuum conductance compared to other designs helps to achieve good transmission, while the spiraling inclined field minimizes beam-spot motion.

TUBE SUPPORTS

Upon hearing of our tube failure, Hyder[2] pointed out that the tension forces along the lower edge of an accelerator tube must be proportional to the cube of its overall length. This would mean that 8-foot FN tandem tubes are much more prone to failure than their 6-foot counterparts in both smaller (EN) and larger (MP) tandems. Hyder and Lionel Fell from Dowlish strongly recommended that the FN tubes be provided with at least one additional support per tube to limit these tension forces in the glue bonds.

Florida State heard of this and soon installed the recommended tube supports in their machine. Our design is a little different from theirs in that we employ a large stainless steel "hook" that resembles a standard HVEC spark shield and is attached to the side of the column in the same way. Two small springs provide a total lift of 50 pounds for 1.0 inch of stretch. The springs attach to the beam tube via a small polished split clamping ring provided by Dowlish. The whole arrangement is easy to implement and causes minimal interference with the column and tube.

REPAIRS

On June 8th we shipped all four tubes to Dowlish in England for evaluation and repair. During shipment #4 tube continued to hold vacuum and only one more glass broke on the other three tubes. We were told that there was little sign of vacuum contamination or electrical damage on the electrodes. (This means that we were either "good" to our tubes or, possibly, not hard enough on them!). Dowlish recommended that all four tubes be completely rebuilt and, with our OK, proceeded to do the job in just over a month.

Upon receipt of the rebuilt tubes in early August we carefully inspected the bonds at every electrode. As shown in Fig. 2, this revealed a number of small bubbles and "crow's feet" within the glue bonds. The latter tended to be concentrated in the same area -- one-third of the way from either end -- where the tube fractures had occured. In some places it appeared that one-half or more of the bond area was filled with many small bubbles.

Extensive discussions with Dowlish ensued. Evidently the features observed resulted from imperfect flatness of the titanium electrodes -- our tubes were the first set with an incline angle as large as 10° and there was, as a result of imperfect manufacturing technique, some residual warpage in the electrodes. Furthermore, since Dowlish does not have a large enough oven to make up an entire tube at once, subsections of about one-third of a tube in length are assembled together with locally applied external heat. This might explain the concentration of visible features near the one-third points.

In due course we received strong formal assurances from Dowlish on the long-term mechanical integrity of the tubes, provided we agreed to utilize the additional supports midway along each tube that Hyder had suggested.

Fig. 2   Enlarged portion of one of the rebuilt Dowlish tubes, showing typical "crow's feet" flaws in the PVA glue bond.

INSTALLATION AND ALIGNMENT

The tube alignment was straight-forward with our usual laser technique, which permits a transverse accuracy of about 10 mils (0.25 mm). We did however encounter some puzzling problems with the longitudinal fit. At first it seemed as though the tubes had come back shorter than they went over, since there was nearly 0.5 inch missing over the length of the machine, and nothing to fill the gap. Eventually we realized that our bellows had become compressed. These are original HVEC formed bellows designed to withstand probably 300 psi, and they are extremely stiff.

We did find that by carefully stretching the bellows we could reverse the set that had occured. The spring constant for these is quite large, about 50 pounds per mil, so we installed bolt rings and used threaded rod for the stretching. If we over-shot the desired stretch by 25% or so while pulling, then the final length came out about right. Some care was needed as well to see that the o-ring surfaces came out parallel. So in effect what was done was to custom fit the various bellows to the gaps in such a way as to compress the bellows absolutely no more than necessary to insert them. This would ensure that when the seals were tightened the bellows would exert the least possible tension on the tube flanges.

The remainder of the installation went smoothly, except for a week or two spent chasing small SF_6 leaks.

REFERENCES

1) Dowlish Developments, Ltd, Dowlish Ford Mills, Ilminster, Somerset, England TA19 OPF.

2) H.R. McK. Hyder (present address: Yale University, New Haven, CT), private communication.

DISCUSSION FOLLOWING J. PREBLE PAPER

LETOURNEL: Okay, thank you, Joe. I think many people will have questions.

ZINKANN: Why is it again you just didn't get new bellows?

PREBLE: Time and money. Mostly time. We were down for a considerable length of time, and we were very anxious to get back on.

WEIL: Did you consider just putting a spacer in to fill the gap, and maybe a little bit over, so you'd run the bellows in compression?

PREBLE: We certainly did. But it's scary introducing any new joints in the tube. We ended up fighting with a pressure leak as it was, which took us a considerable amount of time.

STORM: Can you describe a little bit more about the bellows, how big they are, who makes them? They sound very strange to me.

PREBLE: They are a six-inch diameter stainless steel bellows, part of the original equipment that came with the machine. They're four inches in length. Some have six and some have seven convolutions, and they are just stiff as can be.

MC KAY: What voltage have the tubes been up to since the repair?

PREBLE: The machine has only been up to five million volts. We actually had two leaks which we fixed in succession. We were very, very gunshy of leaks and have been very careful, and it was only yesterday that they even put the five million volts in the machine.

ROWTON: Are these tubes still mounted the same way as HVEC tubes?

PREBLE: They rest on a little arm, on a rubber pad, and they are bolted at the bottom only.

ROWTON: The other question is, and perhaps Lionel ought to answer this, what was the explanation for the breakage?

PREBLE: Well, there are many ideas. There are some problems evidently in the manufacturing of the electrodes themselves. There are some questions about humidity. You have this eight-foot-long tube, and it's going to tend to sag. Those are the things that we're looking at harder.

CHAPMAN: There _is_ a solution to your bellows problem. Like most solutions, it is not particularly cheap, but it's cheap compared to the price of accelerator tubes.

The bellows obtainable now, which are welded diaphragm bellows, are shorter than the HVEC bellows, and they have a very, very low spring content. You can move them with your fingers very easily. So if one has these bellows, you put no force on the ends of the tube at all.

PREBLE: We had talked about buying some of these bellows and still are, but again it was just a matter of time. Delivery time on these bellows was longer than we wished to wait.

FOX: Were the bellows that were originally in the system under considerable tension, or were they as originally mounted?

PREBLE: Well, to be honest, we hadn't really paid any attention. During the last tube installation this problem either wasn't there or wasn't noted.

FOX: The bellows had to be forced apart, or did they have to be compressed? Normally we have to compress our bellows.

PREBLE: Well, since the spring constant is so large, a very small expansion of the bellows produces a large force.

So when you have to get an inner ring and an "O" ring and an outer ring into position, certainly you have to compress the bellows. But then after it's in place, it does have to expand back out into this gap a little. So there is the chance that although you have to compress the bellows to install it, you will put it under tension when you finally clamp it in place.

GRAY: Ken, what is the external working pressure on these diaphragm bellows?

CHAPMAN: It depends from which firm you buy them. I don't have them here, but I have the name of two suppliers. From one supplier the bellows are very much more reasonable. They are in fact rated at considerably less than 100 pounds but in a very short length. They will comfortably stand 100 pounds, because the problem is with them squirming in the long length.

There is another manufacturer that in fact will make them up to two or three hundred pounds, but these bellows do cost on the order of a couple of thousand dollars apiece. They are not cheap.

MC KAY: I am trying to recall when these problems started. My impression is that broken tubes have become a very popular thing in the last five years or so.

Tubes have been around much longer than that. I would like to throw out a question to the floor, asking people if they have any information about tubes falling apart prior to about five years ago.

FELL: My first knowledge of tubes falling apart goes back 20 years. This was a tube that was made for Manchester University, an aluminum tube. It was stored unpressurized for a period of about 10 years. When it was taken out of its box, it was found to be broken.

The repair of the tube was quite simple. It was simply reassembled in the jig, rebaked, and everything went together okay. There was clearly no degradation.

STURBOIS: Do you think that your tubes are under tension now, or do you think you managed to get them installed so that they are in compression?

PREBLE: We worked very hard to make sure they are indeed in compression.

STURBOIS: Did you leave the expanders between the bellows' flanges, or did you just try to stretch them.

PREBLE: We just stretched them out so far that when we installed them they were indeed compressed. We don't have any expanders.

TESMER: As I recall in Derek Storm's and my days as graduate students at the University of Washington, there was a failure of a lens in the terminal on the injector. It was a very violent failure, and I think the glue joint was the failure.

MAN IN AUDIENCE: There was a reason for that having to do with different diameter parts. There was a net pressure from the gas force tearing it apart. That's a little different.

TESMER: Certainly if you pressurize the bellows you cause the bellows to shrink.

MAN IN AUDIENCE: The convolutions are larger in diameter than your end connections. You have a net compression force.

PREBLE: We looked at the geometry of these bellows, and we looked at surface areas. We decided that that just wasn't enough, I mean, when we looked at the pressure in our tank and how much tension we were getting.

JOHN BROOKS: You were up several days before the tubes failed. Were the tubes evacuated during this time when the tank was up?

PREBLE: The tubes themselves were up to air. We were doing

some repairs. That also seems to be possibly a common factor in some of these tube breakages.

LINGREN: I don't understand how leaving the tubes at atmosphere could be harmful, because the manufacturer HVEC ships its tubes at atmospheric pressure, and we sometimes leave one in the box for 10 years before opening it. We have never had the problem of a new tube opening up.

PREBLE: I'm not claiming that this is for sure a reason for tube breakage. All I am stating is that in more than one instance a mysterious tube breakage occurred after the tube had been in use for a prolonged period and then allowed to stay up at atmosphere for several days.

STORM: You said some time ago that you felt there was a correlation between the electrode material and the quality or longevity of the glue joint. Do you want to remark on that?

PREBLE: Well, there are two factors evidently. One, when the electrode is actually manufactured the surface is not perfectly flat. During the manufacture of the tube, considerable pressure and heating is applied, and it tends to flatten out the electrode. After you take that pressure away it likes to restore back to its old shape.

Another is that you will end up with a shear due to temperature cycling up and down, since the glass and titanium will of course expand and shrink at different rates. So you are putting a stress, a shear in that area.

STORM: If you are concerned about trying to return to the cusp shape, it must depend a lot on what kind of material it is. I would expect aluminum would be a lot better than other materials. I don't know what temperature is used. Is it hot enough to melt aluminum?

FELL: The tubes are baked at a temperature of 175°C.

JANSEN: We had a tube break when we first installed it in our machine at Queens in 1966. That was an aluminum electrode tube, and it had been shipped from High Voltage perfectly all right. In fact, the engineers were there when it broke.

It had never been pumped out so far as I know, except for tests at the plant. Also, the machine had never been pressurized at that time. The tube was mounted in the KN machine, cantilevered out from the base end. There was no support at the top end.

We were doing some alignment, we went for lunch and came back and found our alignment was way out. We couldn't understand why this had happened. We began to look for other things, and it turned out a tube had broken.

The failure seemed to be at one of the glue joints, again a third of the way up the tube from the base end, which then somehow stressed the glass, and the glass failed. We never had a satisfactory explanation of why that failure occurred.

So the question is: how do these failures start? Where do they originate? Is it in the glue which then stresses the glass? Is it important how that tube is mounted?

I think in our case it was fairly clear that the failure was one of tension in the top of the tube where the glue failed, and I think we all know that glass is extremely strong in compression and yet is relatively weak in tension.

MC KAY: If you are looking for ancient stories, I seem to remember that in 1950 at the two megavolt electron machine at MIT the column and a tube failed, pretty close to the base. This was a single-end electron machine. The whole thing was lying in the tank. I don't know if this was an HVEC machine or not.

BERNERS: Two of our tubes broke in 1980, and the comment I wanted to make about that was not about the details of the break but that during the rebuilding process I was talking one time to Joe Peoples at Potentials, Incorporated, and he said, "It's very hard now to get good glue. I have the only stock of good glue left in the world, and I don't know what I'm going to do when it's all gone."

I think anybody who really wants to try to pin down the causes for tube breaks should maybe talk to Joe Peoples and find out if it's really possible that there are different grades of polyvinylacetate; if some are accelerating tube grade and some are something else.

KRAUSER: Does anybody know if the tubes are backfilled with nitrogen before shipment, or are they just sent?

PREBLE: Our tubes came evacuated.

MC KAY: The HVEC tubes for our machine had a little filtered band on them, so they were open to atmospheric pressure.

LINGREN: I don't know if most people know this, but High Voltage Engineering has never pulled a vacuum on their tubes after they made them.

They have always just relied on the fact that they know what they're doing. They guarantee them to be leakfree.

There was a problem with this for the last spare tube that we bought. In the laboratory we now have quality assurance control. Everything coming in has to pass quality assurance. We had to write in specs for this tube, that they be evacuated and repacked. Charlie

Goldie agreed to that, that there's no problem, they can do it, but you have to ask for it.

In the years that we have been buying tubes we have only had one that leaked after we installed it. Their record is fairly good.

CHAPMAN: Two quick comments.

Firstly, I believe humidity plays a very big part in the failure of these tubes. Secondly, in answer to Henry Jansen's question, our tube that broke was a stainless steel tube, and when it broke the glass was torn apart.

We borrowed a tube which had been taken in good condition from an accelerator at Argonne, and when we went to pick it up off the floor it was just in pieces. That was an aluminum tube, as distinct from ours, which was a stainless steel electrode tube. In that instance the joints had just come apart neatly.

It appears to be that breakage starts in the seal. If the electrode is truly flat, then it will just come apart. If the electrode was distorted after stamping and then was pressed flat in the tube assembly, when that seal starts to lift it will put forces into the glass and will tear the glass asunder. I'm sure when the tube is cantilevered you get exactly the same effect. As the glue joint starts to lift you put tremendous forces on the glass.

LETOURNEL: Thank you, Ken. I just want to remark that when we changed from 72 inches up to 88 inches we had also a problem which is somewhat similar to this one. We had two thin flanges at the tube ends, and those flanges transmitted forces during bolting. We had a very bad period of time, breaking almost all of the tubes up to the time we went up to 50 millimeter thickness flanges. Of course, it was related to forces transmitted when bolting.

MECHANICAL FAILURE OF ACCELERATOR TUBES

H. R. McK. Hyder* and L. R. Fell+

*A. W. Wright Nuclear Structure Laboratory,

Yale University

New Haven, CT 06511, U.S.A.

+Dowlish Developments Ltd.,

Dowlish Ford Mills, Ilminster, Somerset, England

The mechanical failure of several accelerator tubes in recent years has naturally caused concern among the operators of electrostatic accelerators as to the long term integrity of the glued joints used in glass-metal accelerator tubes.

Failures have occurred in tubes of different manufacture, different electrode material, different design and different mode of use. The proportion of tubes which have failed is small relative to the total number in use. Although all the failed tubes were several years old at the time of failure, there is no obvious single factor which can explain all the failures reported recently. Some failures have also been reported in column structures. These use a different adhesive, different metal laminae and a very different type of construction.

The causes of joint failure in glued glass/metal accelerator tubes may arise in the design, in manufacture, in assembly, in installation, in use and during maintenance. Some of the factors which should be considered are:

1. INTERNAL STRESSES

(a) Because of the differential thermal expansion of glass and metal, the use of thermoplastic adhesive necessarily results in permanent radial stress in the glue film. The manufacturing process is designed to minimize this stress. The mismatch in thermal expansion is greatest in aluminium and least in titanium, but the relative ductility of aluminium may compensate for this to some extent. The plasticity of the glue film is of importance, hence the need to keep the tubes always above 40 F.

(b) As a result of the forming processes, the electrodes are never completely flat and the tubes are therefore subjected to external compression during assembly. When the external force is removed, the electrodes will tend to revert to their original form, applying uneven axial stresses to the glue film and the glass. The effect is more important in stainless steel and titanium because of their greater resilience and because the forming process is more likely to introduce non-planarity.

2.   EXTERNAL STRESSES

Horizontal tubes are subjected to bending stress due to their weight. This stress limits the length of cantilevered tubes and the end load that they can carry. Tubes simply supported at their ends, even those as long as FN tubes, have lower stresses than some cantilevered tubes. Destructive tests on EN aluminium tubes at Oxford showed that for the particular samples studied, there was an adequate safety margin and that the pattern of deflection and failure was as expected. Failure occurred in the glass at the point of maximum stress, and the glue films remained intact. These particular tubes were thirteen years old at the time of test.

The stress due to the weight of the tube can be calculated with confidence. It is sometimes the case that tubes are not simply supported and have additional external loads imposed on them. In horizontal tandem accelerators, the bellows at the tube joints are very stiff and difficult to install. If tube alignment is not perfect the bellows may exert shear forces. If the bellows have been over compressed they may apply an indeterminate tensile force to the ends of the tube. Additional bending moments may be applied by resistor assemblies, spring connectors, or by the column insulators, in the situation where the accelerator tube is laterally stiffer than the column and is used as a structural brace. Other forces may be applied to an accelerator tube when the tank expands with increasing pressure or temperature if the tube supports are not sufficiently compliant.

3. CHEMICAL ATTACK

The glue material commonly used is chosen for combination of properties including high strength, relatively high softening temperature and low water absorption. It suffers only surface attack from water immersion and is claimed to be unaffected by sunlight or age. It is not known whether prolonged exposure to compressed sulfur hexafluoride has any effect on the glue film. Certainly there is no visible evidence of attack and many tubes have operated in SF_6 for very long periods. There is a suggestion that some failures have occurred in relatively high humidity. This may be a contributory factor. Poly-vinyl acetate is soluble in some organic solvents but this is unlikely to occur in service and the primary solvent is driven off during the production process.

4. VIBRATION AND FATIGUE

There is significant vibration in the operating environment both with belt and chain-charging systems, as evidenced by the noise level. It is possible that vibration is a contributory factor in a time-dependent deterioration of the seals.

5. AGE

There are no reports of new tubes breaking. Most failures have occurred in tubes which are at least five years old, some much older. There is therefore a presumption either that glue strengths decrease, or that internal stresses increase with the years. If this is combined with above average internal or external stress then the tube may fail. It is significant that failure occurs preferentially when tubes are not under vacuum, that is when the normal external compressive force has been removed. In one case at least, a tube supported uniformly in a storage box and under negligible external stress was found to have disintegrated after several years of storage at atmospheric pressure.

CONCLUSIONS

There seems to be no convincing reason to doubt the basic glass/metal joint technology. Failure is still relatively unusual. Action should however be directed at reducing all unnecessary stresses, since stress seems to be the major factor in nearly all cases. The manufacturer can play a part by controlling electrode flatness and rigidity and the quality of the glue film. The user should ensure that no unnecessary stresses are applied to the tube by the end support, by the bellows, or by forces appied by the rest of the structure or by spring connections. It may be desirable in some cases to reduce the effect of tube weight on end supported tubes by fitting spring supports at the mid point.

If all these things are done, then the number of future failures may be too low for it to be possible to decide unequivocally which of the several contributory factors is the critical one.

I am grateful to many colleagues, especially John Noé, for comments and information.

MECHANICAL TESTS OF ACCELERATOR TUBES

H.R. McK. Hyder

Nuclear Physics Laboratory, University of Oxford, England

1. INTRODUCTION

Accelerator tubes in horizontal Van de Graaffs are subjected to mechanical forces due to their own weight and to the weight of other components carried by them. In some cases additional forces are applied by tension rods or in the course of maintenance or alignment operations.

Recently a number of instances of mechanical failure have been reported, some without obvious cause. Taking advantage of the availability of a number of old, used tube sections, measurements of the modulus of elasticity and ultimate strength have been made on two 6 ft. long glass/aluminum accelerator tubes recovered from our EN tandem.

We discuss the implications of these tests for the use of similar tubes in horizontal accelerators.

2. MEASUREMENTS

(a) Three point bending tests were carried out on a test machine in the Department of Engineering Science. Both tubes consisted of 74 borosilicate glass insulator rings, cemented to pressed aluminum electrodes with poly-vinyl acetate, with a stainless iron flange at each end. Both tubes were thirteen years old at the time of the tests, had been used for about 10,000 hours and had subsequently been stored under vacuum. The dimensions of the tubes are given in Table 1. The tubes were supported by the end flanges on steel V blocks above a rigid beam. The downward acting piston of the test machine applied a measured force to one of the central insulators through a shaped aluminum block which distributed the load on the glass. The downward deflection of the tube centre was measured with a dial gauge. The downward force was increased in increments of 20kgF up to a maximum of 400kgF for tube C1 and 300kgF for tube C2. Over the measured range, deflection was proportional to force. Tube C1 was 50% stiffer than tube C2, the central deflections being 0.051mm/100kgF for tube C1 and 0.075mm/100kgF for tube C2. No creep was observed when the maximum force was sustained for one hour. The tests were conducted in a dry atmosphere at a temperature of 20°C. The tubes were not evacuated.

(b) End load cantilever tests were carried out by clamping one

end of the tube to a solid plate fixed to the floor of the Nuclear
Physics basement. A horizontal force was then applied via a proving
ring to the top of the vertical tube and the sideways deflection
measured with a dial gauge. The deflections of the base plate and of
the end clamp were measured with an accurate level and with dial
gauges. The arrangement is shown in Figure 1. Using the screw
tensioning device, the sideways force was increased in increments of
~10kgF. During the latter stages of the test, successive increments
were applied at intervals of 30 minutes and both tubes were left
overnight at high stress.

The deflection/force curves for tubes C1 and C2 are shown in
Figs. 2 and 3 respectively. The modulus of elasticity was not
constant in either case, declining to half the initial value at
failure point in tube C1 and to two thirds of the initial value in
C2.

In both tubes, failure occurred some time after a force
increment had been applied. The mode of failure was identical. The
most highly stressed glass insulator failed in tension at the most
highly stressed plane. The insulator which failed remained bonded to
the electrodes above and below throughout the circumference except at
the point where, after the initial fracture of the glass, a peeling
stress was applied to the glue film. Even there, the glue film
failure did not spread and the limited area of separation can be seen
in the photographs (Figs. 4,5). the insulators adjacent to the
failed sections remained intact and undamaged. Over the duration
of the tests (4 days for tube C1), creep was negligible and there was
no measurable permanent deformation.

3. ANALYSIS

3.1 Three Point Bending Test

The deflection of a uniform beam, simply supported at the ends
and loaded at the centre is:

$$y = \frac{1}{48} \frac{Wl^3}{EI}$$

(see e.g. R.J. Roark, Formulas for Stress and Strain: McGraw Hill
1965).
where W is the load, E the modulus of elasticity, I the moment of
inertia and l the length between supports. Because of the difficulty
of allowing for the non-uniform composite structure of the tubes and
the effect of the semi-plastic glue films, we compare the
measurements with calculations based on a continuous glass tube
having the same I.D. and O.D. as the actual insulators at the point
of minimum cross section.

The resulting values of E, the modulus of elasticity are as
follows:

Tube	C1	C2
$I(= \frac{\pi}{4}(r_2^4 - r_1^4))(mm^4)$	5.67 X 10^7	4.09 X 10^7
$1(mm)$	1905	1905
y/W (mm/kgF)	.00051	.00077
E (kgF/mm^2)	4981	4573
E (GN/m^2)	48.9	44.9

The quoted value of E for borosilicate glass is $68GN/m^2$.

3.2 Cantilever Bending Tests

The deflection of a cantilever, fixed rigidly at one end and loaded at the other is:

$$y = \frac{1}{3} \frac{Wl^3}{EI} \quad \text{(see e.g. R.J. Roark)}$$

Making the same assumptions as before, the modulus of elasticity can be derived from the slope of the stress/strain curve at zero stress and just below breaking stress.

The results are:

	Tube	C1	C2
	$I(mm^4)$	5.67 X 10^7	4.09 X 10^7
	$1(mm)$	1905	1905
At zero stress	dy_0/dW_0(mm/kgF)	0.0127	0.0149
	$E_0(kgF/mm^2)$	3199	3781
	$E_0(GN/m^2)$	31.4	37.1
At breaking stress	dy_B/dW_B(mm/kgF)	0.0256	0.0237
	$E_b(kgF/mm^2)$	1587	2377
	$E_b(GN/n^2)$	15.6	23.3

3.3 Breaking Stress

The applied moment M which caused tube failure was 350kgF X 1.905m (6541Nm.) for tube C1 and 220kgF X 1.905m (4111Nm.) for tube C2.

164

The maximum stress in the glass at breaking point is given by

$$P_{max} = \frac{r_2}{I} M_{max} = \frac{4r_2}{\pi(r_2^4 - r_2^4)} M_{max}$$

Hence, since $r_2 = 0.1075m$ (both tubes), we have:

$$P_{max}(C1) = \frac{.1075}{5.67 \times 10^{-5}} \times 6541N/m^2$$

$$= 12.4MN/m^2 \ (1.26kgF/mm^2 \text{ or } 1795 \text{ p.s.i.})$$

$$P_{max}(C2) = \frac{.1075}{4.09 \times 10^{-5}} \times 4111N/m^2$$

$$= 10.8MN/m^2 \ (1.10kgF/mm^2 \text{ or } 1562 \text{ p.s.i.})$$

4. CONCLUSIONS

Two accelerator tubes of similar overall dimensions and construction have been tested to destruction. Both tubes failed some time after the slow application of a bending movement. Failure occurred at the point of maximum tensile stress in a glass insulator. None of the glued joints failed. The ultimate stress in the glass was $11.6MN/m^2 \pm 7\%$.

Assuming that the composite structure of the tube is equivalent to a homogenous glass tube of cross section equal to the smallest part of each insulator, the three point bending tests yielded values of the modulus of elasticity at low stress of $45-50GN/m^2$. This is 30% lower than the quoted value for borosilicate glass.

The cantilever bending tests were less consistent. At low stress the values ranged from $31-37GN/m^2$ and at breaking stress from $\sim15-23GN/m^2$. If the safe working stress is taken as one fifth of the ultimate stress, the safe limit for strain at the unsupported end of a standard 6 ft. X 8 3/4" accelerator tube is only 1mm. Taking the weight of tube C1 as 79kgs and that of tube C2 as 72kgs, the additional safe load which can be applied to the end of a horizontal cantilevered tube is 30kg (C1) or 8kg (C2).

By contrast, a similar tube which is simply supported at the ends can safely sustain a central load of 240kgs (C1) or 140kgs (C2).

5. ACKNOWLEDGEMENTS

I am indebted to Dr. C. Ruiz for making available the test equipment in the Department of Engineering Science and to Mr. R. Stone for his assistance with those measurements. Dr. Z.Y. Guo and Mr. A.B. Knox were responsible for taking much of the data and for analysis and I am most grateful to them for their help.

TABLE I

Dimensions of accelerator tubes C1 and C2

		C1	C2
Overall length		1.99m	
Insulated length		1.88m	
Number of insulators		74	
Number of electrodes		75	
Adhesive		Poly-vinyl acetate	
Electrode:	material	aluminium	
	outside diameter	222.5mm	
	thickness	2.2mm	
Insulators:	material	Borosilicate glass	
	outside diameter	216mm	
		C1	C2
	inside diameter at glue line	177mm	181mm
	minimum inside diameter	163mm	181mm
	maximum inside diameter	177mm	191mm
Weight		79.3kgs	72.1kgs

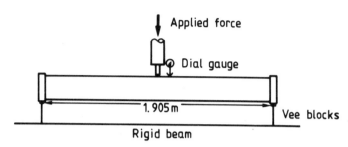

A. THREE POINT BENDING TEST

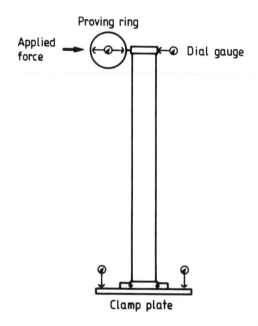

B. CANTILEVER BENDING TEST

FIG 1

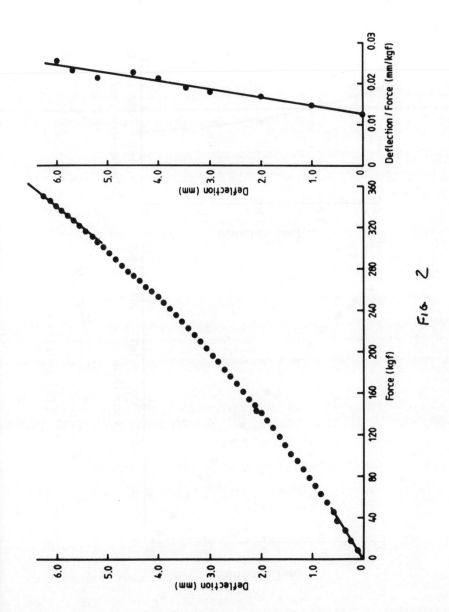

Fig 2

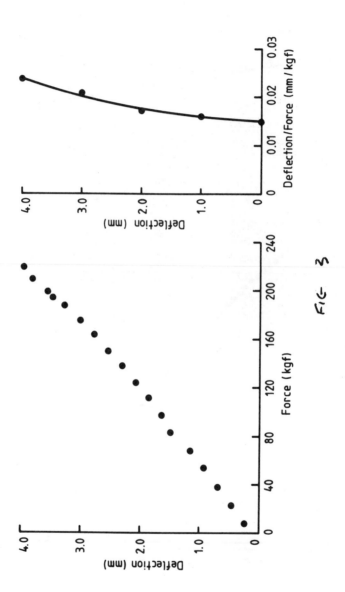

FIG 3

DISCUSSION FOLLOWING H.R. McK. HYDER PAPER

LETOURNEL: Any questions? Please.

BERNERS: Dick, I would like to offer one change to you tabulation of failures.

At Notre Dame the tubes you have listed as not being under stress, but they were mounted in the machine, only at the ends.

HYDER: My apologies. If anybody has taken careful note of that, perhaps they would correct the entry under Notre Dame.

BERNERS: I have a question about the failure of the bellows. If the bellows do relieve their own internal stresses by settling into a different shape than they originally had, then it's hard for me to understand why tension applied by the bellows would cause failure after the assembly had been in place for a number of years. I would think the likelihood of failure goes down with time.

HYDER: One point I omitted to make was that during those tests I carried out on the aluminum tubes at Oxford, failure did not occur at the moment of increasing the stress. During the failure tests I racked the force up by a few tens of kilograms every hour or so. The failures on both tubes occurred some tens of minutes after the last increase in force.

So I think there are some time-dependent effects going on here. It doesn't altogether answer the question.

As to the long-term stress in the bellows, I don't really know how that relaxes with time.

PREBLE: In our bellows we found that we required four to one, if we wanted to realize some change in length, we had to stretch if four times that.

After you could initially stretch it and let it back, you would get some change. We did leave it stretched for not a long time but an hour, two hours, three hours, so it's not a wonderful test, but we did realize that effect over momentary stress.

FOX: In your test of the tube that was horizontal how long a period did the test take?

HYDER: I had one of the two tube sections sitting there under the maximum load for about four days.

FOX: Was there any perceptible yield?

HYDER: No. No creep that I could detect. That's rather a short period. On the other hand, the force was pretty enormous, and this was the equivalent of about four people sitting on the middle of an EN tube that is a big force.

STARK: All of these tubes have been used tubes. Is there any thought to radiation damage to the plastic that may make it more hygroscopic or something?

HYDER: I would have thought that was most unlikely, because the radiation levels in these tubes are quite small. Nearly all the ones I've been talking about today are suppressed, so the radiation levels are quite low.

If radiation damage was serious I would have thought that the horizontal electron machines where the tubes go black within a few hours and the radiation doses are thousands of RADS would be much more likely to fail.

I'd very much doubt whether radiation plays any part in this. The glass isn't even brown in most of these tubes.

MC NAUGHT: The thought occurred to me while you were talking, both you and some others have indicated humidity may be a contributing factor. It seems to me in recent years some labs have been advocating washing down their machines during a maintenance period. Maybe this isn't a good thing to do, or would that be over such a short period of time it doesn't matter?

HYDER: I don't honestly know the answer to that. We did in fact wash the EN in tandem at Oxford twice. I think that the tubes I destroyed were not in the machine at the time we did the washing, so I don't really have any evidence on that.

EVANS: In these calculations are you also adding the tank expansion as a function of the pressurization? Does that apply any lateral stresses to the internal tubes?

HYDER: It will tend to do that, because the tank expansion is meant to be taken up by the bellows. The bellows is so stiff that as the tank expands the bellows will presumably pull on the whole assembly. So that will add a bit.

An EN tank I think expands an eighth of an inch or thereabouts, so there are six bellows, so the amount is small compared with movement of the bellows themselves, but I think it will add a bit.

STURBOIS: Have you given any thought to just straight loading the bellows' flanges, so that the tubes are always in compression without them changing when you let it up to air?

HYDER: I think the problem with the EN and the FN is that you need very strong springs, and there's very little space. You see, the back of the bellows is right up against the column structure. One of the difficulties of the whole exercise in fitting the bellows is that it's so hard to get around to the back and to see what you're doing there's very little space. But it's a thought.

SOUTHON: One quality control problem was how clean the parts are.

If it's not proprietary information maybe you could tell us what is done to clean the parts before assembly.

FELL: We haven't been aware of any problems in climate tests. The techniques are relatively simple. We use solvents and a baking technique which effectively dries out any solvent residue or moisture that might be in the components.

Also in the course of preparation of the glue fillings the components are baked again to dry off solvents from the glue.

Obviously all precautions are taken to ensure there is no contamination from the time the components are cleaned up to the time they are to be assembled into a tube.

Furthermore, in the baking process or actually assembling the tubes there is another effective cleaning process by temperature.

Finally, when we do a vacuum test we take a mass spectrometer, and any contamination in the tubes would certainly show up there. We have never seen any contamination in the tubes.

LARSON: I'd like to respond to what Ed Berners asked, about why the Stony Brook tubes didn't fail immediately if it were damage from the bellows.

It seems to me there's irrefutable evidence of aging effects occurring when you've got tubes that literally fall apart in several places simultaneously. That's got to be aging. I can imagine a single failure could come from any kind of source, but multiple failures in the same tube have just got to be from aging, and there fore since presumably the people at Stony Brook weren't alert to the possibility of tension from the bellows being applied to the tubes it could well be that when the tubes were new and in good condition they were put in tension by the bellows, and only after they had aged substantially did this make a difference.

BERNERS: If that's the case, then aging perhaps is something that occurs in polyvinylacetate, no matter what environment it's in, because the Stony Brook tubes were aged in the machine.

HYDER: Well, aging alone can't explain things, because these tubes which I tested had been aged for 13 years and were pretty strong, so there has to be more than one factor involved.

In the case of Stony Brook the aging was only seven years, and there are many other tubes which have gone for twice that length of time, but it's possible that the stress applied for those seven years was somewhat higher than in other machines because of the addition of the weight stress in the long eight-foot FN tubes, plus an additional stress from the bellows.

. LARSON: I have a question for Dick. Can you offer any explanation of why with the Stony Brook tubes, several tubes failed consistently in one place, and that's not in the center of the tube itself?

HYDER: No, I think that there is a bit of a puzzle there. It doesn't seem to be related to joints. It's in a region where the stress is high but not at a maximum.

It is not I think related to the orientation of the inclined fields. That was another thing we looked for, to see whether the titanium had all warped in the same way. I tend to think that this number of 32 plus or minus two may actually be a coincidence, that the individual electrodes we know were not as flat as they should have been, and it just happened that the worst of them were at these points. The fact that it was the same point or close to the same point in the three tubes, I can't see any common factor there really.

PREBLE: It's just really surprising that it would happen, same place, three tubes, at the same time.

HYDER: Well, I think the significant factor there is that the tubes must clearly have been weak. The external compressive forces from vacuum over tank gas were removed, so they were then put under an increasing stress. They were all quite consistent. If one went, there was a high probability another would go. I think that's the reason why the three of them failed at the same time.

LARSON: In follow-up, is it possible that something about the mounting could actually have transferred the maximum stress from the center of the tube, which is a gravitationally applied stress to slightly one side, something to do with either the bellows or some other force.

HYDER: As Joe Preble pointed out, the mounts themselves have got some rubber pads underneath, and it's difficult to see them transmitting much force to the tube, although there are screw fixings. These things aren't really very rigid, so I doubt whether they are actually imposing any peculiar and abnormal stress on the tubes.

LEIDICH: Leidich. It seems to me as an outsider, because we haven't had any problems with tubes, they were lying around for years then we just picked up and put them in and they work. Is the problem, maybe there are two categories, one, Mother Nature breaking the tubes, but aren't there a lot of column two problems in the charging system conversions, and it might be vibration?

HYDER: That's a possible factor, but of those aluminum tube failures, most of these are things which have been sitting outside accelerators, with no vibration at all. So I don't know.

The Stony Brook system, I think one can explain that without recourse to vibration, and the FSU one I don't know, but you are now down to one case basically where vibrations might have affected those.

CHAPMAN: We have the original bellows system and have not converted to the Peletron, so there's no charging system change.

PREBLE: We also monitor vibration in the machine very carefully, because we do worry about the Laddertron creating excessive vibration in there, and we saw no signs of a lot of vibration.

STIER: I was wondering how parallel surfaces were when you first noticed the difference in the lengths of the tube and the difference in distance from the tube flange surface to the bellows surface. If the bellows are that strong and they are not perfectly parallel it sound to me like they will torque on the end of the tube like crazy.

PREBLE: We measured a maximum difference of .090 inches.

It's a six-inch diameter, four-inch long bellows, with different lengths varying by .090 around the circumferences of the bellows. That was the maximum difference.

CHAPMAN: I just wanted to ask what you feelings were about the necessity or desirability of supporting the tube at the center of the span.

HYDER: I'm not certain about the necessity, but we have had two failures. We don't know to what extent the joints in those particular tubes may have been weaker than the average.

If it's a fairly simple matter to provide the support at the center, you can reduce the stress in the glass by a factor of three or four with very little trouble. It seems to me that that's an insurance which can be taken out for a few hours work and a few tens of dollars and may well be worth doing. You can't say the tube will necessarily fail if you don't do it, because lots of FNs of course have tubes in them which are running happily without it.

But, on the other hand, it does seem to me that that may be worth doing. It's simple to do and doesn't cost very much.

CHAPMAN: In this particular respect the FN tube is particularly vulnerable because of its length, length to diameter, isn't it?

HYDER: I certainly think it's an insurance worth taking out until such time as one can replace bellows with something which is more compliant.

MILLS: Is the bellows a single-wall, double-wall, or triple-wall bellows? You can increase the elasticity of the bellows and maintain pressure holding ability by using a multi-plied bellows.

HYDER: The edge-welded bellows I think of 200 pounds have to be multi-ply, then once you get two plies into these things the permissible movement is reduced, and they become quite stiff.

The convoluted bellows, I'm not certain. Does Frank Chmara know whether those are single-wall?

CHMARA: No.

HYDER: I think they're probably single-wall and fairly thick walls.

MILLS: They will flex more easily if you use triple-wall bellows. You can allow greater elasticity and still maintain pressure by using multi-layer, convoluted bellows.

LETOURNEL: I would like to comment on that. We were obliged in the process of upgrading our MP to go from 72 to 88 inches HVC tube. I mentioned we were obliged to go from 32 millimeters to 50 millimeters for the flange thickness, and at this time we have to, have this type of bellows, and we were obliged to squeeze it in a very small location, so the bellow transmitted forces on the tube from the compression.

WEIL: We have a 20-year-old CN accelerating tube, aluminum electrode, virgin, never been out of the box. I'm a little bit afraid to look in the box after hearing all the discussion.

However, we are in the process of purchasing a new tube, and if anyone is interested, if it is still together, we probably will have it available for sale.

RICE: We have three old, retired EN tubes that are in crates.

Since it was reported they fall apart, before each SNEAP I open them up and lift one end. They are intact and being stored in southern California, which is relatively dry.

Perhaps that's useful information. To anyone who would like to store their tubes in my vault, I have plenty of room.

MAN IN AUDIENCE: Are they under vacuum?

RICE: I have never opened them. They have been in crates for more than 10 years now. There is some fitting on the ends which would allow you to evacuate them. I do not know what's inside.

TESMER: Since these tubes spend a great deal of time in high pressure gas, is there any possibility of any permeability of the gas into the plastic, such that there is when you take them out some sort of stress like forming bubbles? In some materials you can have injected gas bubbles. It's a very far-fetched idea.

HYDER: I think all one can say is there is no evidence of SF-6 permeating through the seals. There is no visual evidence of any thing happening. At some microscopic level there may be some diffusion of SF-6 into the seals. I think one really doesn't know whether this plays any part. I don't believe it plays a major part. Other wise we would have seen a catastrophic difference between SF-6 machines and nitrogen CO_2. I don't think that's the case.

CHAPMAN: One last comment on that point. Charlie Goldie from High Voltage when they repaired some tubes by rebaking the tubes claimed that they had a certain amount of trouble with bubbles forming in the vinyl after they sealed, which they claimed they felt was due to the pressure gas from the accelerator having been absorbed in the vinyl acetate and coming out again in the baking process, and they were in fact rather reluctant to just rebake the tubes. They wanted after that to take the tubes apart after any length of time in the accelerator. There may be some penetration of gas into the vinyl acetate.

PREBLE: Just a comment. I also have some pictures of broken tubes, if anybody is interested. You can see what we were looking at.

MAN IN AUDIENCE: We have a PN machine that was installed in 1970, and in '69 the machine was converted to positive, and we had a tube which was aluminum. We changed to stainless steel.

Engineer from High Voltage did the installation for us, and he just put the aluminum tube in the crate that it came in.

At that time we were not ready to use the machine as a positive machine, so I changed it, within six months I changed it back to aluminum and put back the stainless steel tube in the crate. In 1982 we reconverted the machine back to positive, and the aluminum tube went back in the crate, and since then it has been in the crate. The ends were capped, but no other precaution has been taken.

At this time we have no intention of going back to the electron mode, but we don't want to discard the tube, because at some time it may go back. All I want to know from the floor is if there is any precaution to be taken.

HYDER: I don't believe it will do any harm to evacuate the tube and leave it under vacuum.

LETOURNEL: Any other comments or questions? So now we close the session.

SESSION VIII

OPTICAL PUMPING IN THE HEIDENBERG POLARIZED HEAVY ION SOURCE (PSI)

H. JAENSCH

At the Max-Planck-Institute fur Kernphysik in Heidelberg, we
have an MP Tandem and the usual running voltages are around eleven to
twelve megavolts. A normal conducting post accelerator is used with
that. For this machine, we have built a polarized ion source for
lithium and sodium in succession of one that had been used at the EN
Tandem. One of the most important new features in the source was the
use of optical pumping in the production of polarized heavy ions:
So, I will say something about the optical pumping mechanism,
something about what the Doppler effect does in this field,
application to a source, certainly how does it work in that source,
something about RF transitions which you always need in polarized ion
sources, you have probably heard about them a couple of times so here
again there are some, and then I want to spend some time on an
improvement, which is going on on this already improved ion source,
which I will hopefully show you, that it is.

Optical pumping is a very simple process. In general, you just
irradiate your sample of atoms and induce changes of some level
population.

An important scheme for polarized ion sources, is that you make
use of the selection rules of an optical excitation. You induce only
transitions which increase the angular momentum quantum number by
one. This is done by circularly polarized light in a small magnetic
field parallel to the laser. In the excitation you always gain an
angular momentum quantum, which you might or might not loose in the
spontaneous decay. After several cycles the atom ends in the highest
angular momentum state and is thus highly polarized. Part of the
atoms may also be lost in a not pumped state. The process is
illustrated in Fig. 1. The loss of the unpumped state F = 1 for Na
is large. Only 24% of the initial population end up in the state
12.2>. The rest is defocused in a 4-pole Stern-Gerlach magnet.

You cannot do the same scheme with radio frequency means because
in the radio frequency regime, the lifetime of the species you deal
with is so long that you don't have a redistribution as you have for
the optical transitions by spontaneous decay.

In the polarizing optical pumping you want to pump all atoms not
only a small velocity bin, thus one has to worry about Doppler
broadening of the optical line. This broadening is overcome almost
automatically by using perpendicular geometry. When the laser hits
the atomic beam at 90 degree incidence, the first order Doppler
effect vanishes. The second order one can be neglected. The angular

spread of the atomic beam controls the effective linewidth and it has to be chosen so, that the laser can still cover the complete beam.

Now we look at a schematic view (Fig. 2) of the ion source as we use it. There, is an oven, that is basically a steel can, where sodium or lithium is put into, and an orifice. Then we have two apertures, which are heated and take away most of the beam. Next we have an interaction region with a laser. This interaction region is magnetically shielded from the following quadrupole magnet. This Stern-Gerlach takes out part of the beam which we, with the present method, cannot transfer into the state that we want it in. Then we have an RF transition and an ionizer. The ionizer is in a large magnetic field. It's just a hot tungsten strip which is oxidized. By surface ionization, we reach an ionization efficiency of close to one hundred (100) percent. That is very favorable for the alkaline atoms. Then behind the ionizer, we have a second laser which analyzes the atomic beam state distribution. This is necessary because the things you do up front, the optical pumping and the RF transition, have to be well adjusted and these adjustments have to be well controlled. The ion beam is extracted from the tungsten surface and bent into a charge exchange region, where we go from positive to negative ions. This is done in a magnetic field, so we don't loose the nuclear polarization. Then there is a spin procession system, a so called wien filter, where we can, in a controlled way, rotate the spin in any direction we want to.

In Fig. 2 (a) we see the result of the optical pumping and the cleaning process of the Stern-Gerlach magnet. The spectra shown are recorded by the laser induced fluorescence (LIF) technique. For this the ionizer is removed pneumatically (Fig. 2) and the diagnostic laser activated. We tune the wavelength of this laser from just below the Dl fluorescence lines to just above it. The fluorescence is recorded by a phototube. Whenever an atom in a certain state is present we record its fluorescence when the laser reaches the excitation wavelength. If no atom is present there is no light. This method allows to measure the relative occupation of all hyperfine magnetic sublevel and thus gives complete information on the atomic and nuclear polarization.

Analysis of trace (a) and Fig. 3 shows that 90-95% of the atoms behind the 4-pole are in state 1 (on atomic state 1 $F = 2$ $M_F = 2>$ or nuclear magnetic level $M_I = 3/2$.) But one should keep in mind that this is only 24% of the initial atomic current, the rest is deflected in the 4-pole. A note should be added on the conditions of the optical pumping. At the platform the lase intensity was about $40\text{-}80 mW/cm^2$, which is easily obtainable from a commercial Dye laser, with the dye R6G for Na and DCM for Li. The magnetic field in the interaction region was about 5 Gauss and has to be sufficiently homogeneous. The stray field of the 4-pole (Fig. 2) was greatly reduced by two soft iron plates. The pumping region itself is inside a soft iron tube, in which the 5 Gauss field is generated by 2 tiny coils in Helmholtz geometry. The laser runs along the bore of the tube and for the atomic beam and a viewport suitable holes are provided. The magnetic shielding of the two plates was so good, that in their central bore a small coil had to be installed providing a magnetic field parallel to the atomic beam to avoid depolarizing zero field crossings.

As we saw in Fig. 3 (a), we can produce an atomic beam in a single magnetic substate. For a nuclear physics experiment it is highly desirable to prepare all nuclear spin states (I = 3/2 for ^{23}Na and ^{7}Li; I = 1 for ^{6}Li). Fast switching (seconds) is advantageous. To achieve this we employ medium field transition (MFT). This radio frequency device works similar to a weak field transition (WFT), which are used in most 'atomic beam' polarized ion sources. The MFT runs at 50 MHz for Na (30 MHz for ^{7}Li) and is thus in the nonlinear part of the Breit-Rabi diagram. The rf power and the gradient field in the device are constantly running. Switching from state 1 to either states 2, 3 or 4 is achieved by varying the average magnetic field in the transition region (for state labels see caption of Fig. 3). The result of this transition is shown in Fig. 3 traces (b) to (d) corresponding to the beam in nuclear magnetic substates M_I = 1/2 to M_I = -3/2. Thus the goal is reached to provide a beam in any nuclear magnetic substate. From this pure polarization tensor moments t_{10}, t_{20} and t_{30} can be constructed by suitable time averaging.

The currents in μA, achieved with this setup are for Na (Li) positive extracted current at the tungsten ionizer 40 (15) and negative current at the source exit 0.8 (0.3). We inject roughly one half of this into the accelerator and get for Na in the chargestate 9+ an analyzed current of 60 nA. The Na polarization has dropped to about half the injected value due to beam foil interaction in the terminal stripper. Without further loss of polarization beams produced at 110 to 120 MeV have been decelerated to 48 MeV and accelerated to 184 MeV. The analyzed Li current was 200 nA in the 3+ state at 44 Mev.

Now I want to discuss recent developments of the described scheme. The goal is to increase the intensity of the source by a factor 4 by utilizing the 3/4 of the atomic beam defocused in the 4-pole magnet. The problem is that the laser can only pump one of the alkali ground state multiplets (Fig. 1). The aim is to create two frequencies out of what your dye laser gives you and to use these two to pump both ground state levels. This was achieved using a lithium tantalate (Li Ta O$_3$) crystal in an electro-optic modulation (EOM) device. In this device the laser is phase-modulated at 886 MHz, which is half the Na hyperfine splitting. The splitting of the first sidebands is then 1772 MHz corresponding to the Na ground state splitting. The dimensions of the crystal were 0.65 x 0.5 x 25 mm^3.

The z-direction was perpendicular on the metalized 0.5 x 25 mm² face.
Ten watts of rf power were sufficient to reach maximum laser intensity
of about 35% in each of the first side bands. Fig. 4 illustrates
this pumping scene. We were able to transfer 97% of the total atomic
beam into the desired state 1 (M_I = 3/2). Switching of this
spinstate with the MFT was possible as well, so that the LIF picture
was very close to Fig. 3, just slightly better. Usage of the device
in a beam time has to be awaited. The EOM has turned out to be a
very broad band device (200 to 1800 MHz), so that usage at ^{6},^{7}Li and
^{23}Na is possible. One should note, that usage of such a device with
a variable frequency can make the second dye laser for LIF in our
scheme obsolete. Its role would be taken by a small portion of the
pump laser appropriately offset by the variable EOM.

Another completely different development of the source is the
reconstruction of the Cs charge exchange unit to increase the
magnetic field from 0.1 to 0.25 Tesla for better hyperfine
decoupling. At the same time the reliability of operation (Cs
filling procedure) should be increased.

Finally, I want to compare the figure of merit, which is
Intensity x (Polarization squared) I x P^2 for a polarized ion source.
For the different source schemes this is done in Fig. 5. Clearly the
source utilizing optical pumping together with modulation technique
is greatly superior.

FIGURE CAPTIONS

Fig. 1: Scheme for optical pumping with one laser for ^{23}Na or ^{7}Li.
         The ground state (F = 1.2) and excited state (F^1 = 1.2)
         splitting of the D1 line are shown (vertically) along with
         the corresponding m-states (horizontally). Final state
         occupation after several pump cycles is 24% in the F = 2,
         M_F = 2 and 76% in the F = 1 multiplet. The shaded area
         includes the states defocused by the 4-pole magnet.

Fig. 2: As in the figure caption: Schematic...

Fig. 3: As in the figure caption: LIF spectra...

Fig. 4: Scheme for optical pumping using an electro-optic modular
         (EOM). The laser is turned halfway between the F = 2 and
         F = 1 excitation lines. The EOM produces two strong first
         sidebands which are used to pump all the atoms of the F = 2
         and F = 1 multiplet. The final state distribution after
         several pump cycles is 100% of the atoms in state F = 2,
         M_F = 2.

Fig. 5.   Table giving the figure of merit for the polarized ion
sources Intensity (I) x Polarization (P) squared for the
different source schemes. 'Conventional' employing only a
Stern-Gerlach magnet and rf-transitions; 'optical pumping
conventional' using optical pumping and a Stern-Gerlach
magnet to remove unwanted beam components as well as a
medium field transition (MFT); 'optical pumping and
modulation' utilizing only a modulated laser beam for
optical pumping and an MFT.

184

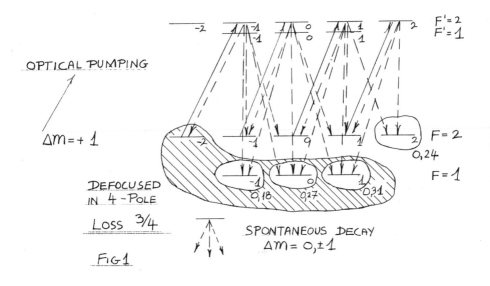

OPTICAL PUMPING

$\Delta m = +1$

DEFOCUSED
IN 4-POLE

LOSS $3/4$

FIG 1

SPONTANEOUS DECAY
$\Delta m = 0, \pm 1$

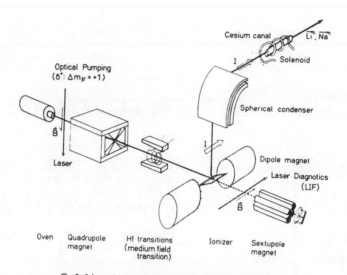

Cesium canal

$\overrightarrow{Li^+}$, $\overrightarrow{Na^+}$

Solenoid

I

Optical Pumping
(δ^+: $\Delta m_F = +1$)

Spherical condenser

$\vec{B}$

Laser

I

Dipole magnet

Laser Diagnotics
(LIF)

$\vec{B}$

Oven   Quadrupole   Hf transitions   Ionizer   Sextupole
        magnet   (medium field   magnet
                      transition)

Fig 2 Schematic view of the source for polarized heavy ions, "PSI".

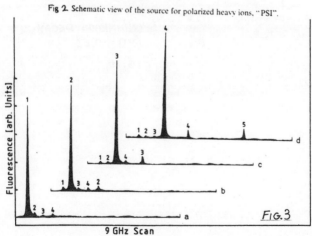

FIG.3

Fluorescence [arb. Units]

9 GHz Scan

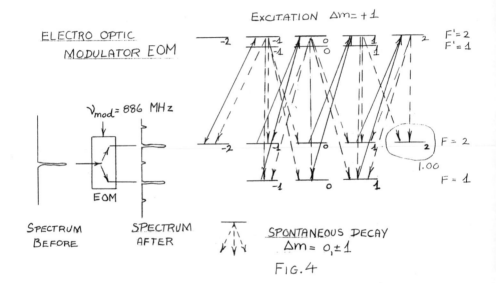

ELECTRO OPTIC
MODULATOR EOM

EXCITATION Δm = +1

ν_{mod} = 886 MHz

EOM

SPECTRUM
BEFORE

SPECTRUM
AFTER

SPONTANEOUS DECAY
Δm = 0, ±1

FIG. 4

SOURCE	I	P	$I \times P^2$
CONVENT.	0·5	0·5	0·125
OPT. PUMP. CONVENT.	0·25	1	0·25
OPT. PUMP. + MODULATION	1	1	1

$$FIG\ 5$$

DISCUSSION

H. Jaensch

MR. ADAMS: Your name, rank, and serial number with Fortissimo if there are any questions, please.

MC KAY: It looks like a vary elegant source. What would the rough cost be to put together such a source?

MR. HEINZ JAENSCH: If you put it together from scratch and don't have all the supplies, in our case we had almost all power supplies, because we had the old source. I would estimate if you really have to buy everything including things you order in the workshop and they make it for you, roughly one million, with the laser system and cooling and high voltage and all that stuff. This is probably a fair number.

The number quoted in the reports which certainly has never been published but internally it's much, much lower.

LARSON: Dollars?

MR. JAENSCH: The laser system is about 400,000 Deutchmarks, and it may be a bit more expensive now. Well, not necessarily. And, then there are the power supplies and the platform and cooling.

ION SOURCE WORK AT SIMON FRASER - I. IYER
(Transcribed from the tape of this session)

We are developing a sputter source at MacMaster. I don't really have any results to show you because we are still in the initial stages of testing it but I do have a cross sectional diagram of the source. It looks like Middleton's source though we do make our source with a spherical ionizer. That is the spherical ionizer there and we have cesium loaded here. The cesium vapors enter this ionizer assembly from there and this piece here, that stands on four legs, has holes in it from where the cesium enters and it falls onto the ionizer. Because of the radial geometry, it is focused on the sputter cathode and we have a negative ion beam coming off from there. It looks like Middleton's spherical ionizer, we don't make this source, but the difference is that we are putting the whole extraction potential on the cathode here, and a hole of minus 30 killovolts there.

We have a prototype and so this is what the cathode looks like in the prototype, but we plan to have a multi sample wheel. Ultimately it should look like that, like Ken

Chapman's inverted sputter source with a multi sample wheel. Now, that is what the source looks like.

We tried to get a beam out of it a couple of times. So far, we haven't got a very good beam we could just get up to a couple of microamps. We feel that the problem is with alignment and we would like to improve the concentricity and we hope to get better results.

191

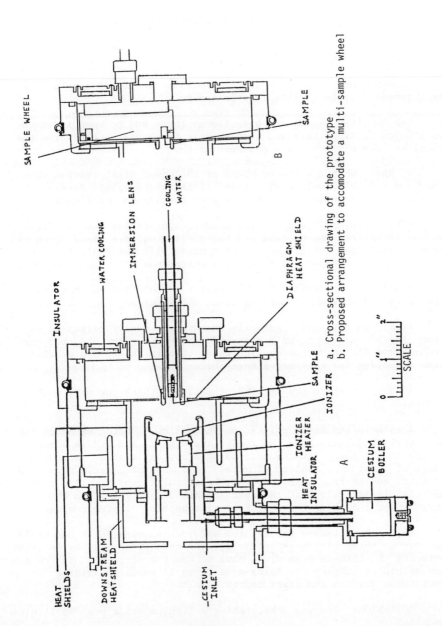

a. Cross-sectional drawing of the prototype
b. Proposed arrangement to accomodate a multi-sample wheel

Discussion Following I. Iyer Paper and
General Discussion on Ion Sources,
LED Replacements, Suppliers and Belts

ADAMS: Any questions.

LARSON: How do you remove the wheel in your multi-sample
arrangement? It looks like it's bolted in.

INDIRA IYER: We have a back flange there and to open it there
is a Cajon fitting there and when we have to change samples from the
wheel we have a hinge there and open it up.

MC KAY: What are the problems of the power dissipitation when
you have a 30 killivolt cesium beam hitting the sample? Have you
taken extra precautions to get very good cooling on the sample?

IYER: Yes, in the prototype that we are working with at first
we had in between the cathode and the cooling water a piece which was
made of stainless steel. We were not happy with the cooling, so we
changed that to copper. I think it's better now.

STURBOIS: Sturbois, Ohio University. Would you put the other
sketch up and show where cooling is on this source?

IYER: In this one. Actually we have not really talked so much
about it, because we haven't even made the wheel right now, but most
probably we will have the cooling going through there, and we also
have a cooling for the back flange there.

ADAMS: If there are no other questions, thank you very much,
Indira.

Are there any other points you would like to bring up about ion
sourcery? I usually have the recipe every year, and I have a new
recipe for making a lithium beam using a sputter source. We put a
piece of lithium on a piece of metal, take a Bernz-o-matic torch and
fire it, and it really goes up in a beautiful flash. We load that
material immediately into the sputter cathode and put it in the
source, and run about three or four microamps of lithium current.

We had been taking the lithium and just packing it in and would
get two or three microamps initially, and then after that oxide layer
wore off the lithium beam drops down to about a microamp. But this
new method of putting the Bernz-a-matic to it seems to oxidize
everything just to the right degree.

MONSANTOS: Are you accelerating a lithium oxide beam or lithium
hydride?

ADAMS: No, we are accelerating a lithium beam.

WILL: Just before we came I was talking to Dave Hopkins, our ion source engineer, and he had tried lithium chloride and had something on the order of 100 nanoamps, and then he tried lithium oxide packed in silver, four volumes of silver powder to one of lithium.

With an 860 Milton type sputter source, and four parts silver, one part lithium oxide, packed with not much care to keep it dry, he got 20 microamps of lithium, he didn't accelerate it. That is quite a bit, Lithium oxide seems to be part of the secret.

SOUTHON: Relative to this business of using a compound to obtain a beam, for example a lithium compound for lithium, we found in cases like that you use the most simple-minded chemistry in picking out which compound you should use. You don't use a halide, because with lithium is least likely to come out negative. You use something that's bonded, like the oxide.

In all the cases where we have looked at the most simple-minded chemistry seems to work.

BERNERS: I have a question for the people from Seattle, Washington. You said in your lab report that you have done extensive modifications to your 860 source in order to improve the energy range of the ions and also to improve reliability. I'd like to have details about what those modifications were.

DEREK STORM: I can tell you a little. I am speaking somewhat from hearsay, but I understand that when he got that source the way the extraction voltage and the focus voltages were connected, the energy of the ion coming out was the sum of these two voltages.

He felt that would make it difficult to tune the beam, because the beam energy would be changing at the same time as the focus. He modified it so that one power supply supplied the full high energy, and there was a focus supply. He did some other things to it, also.

I think he went through the usual cycle of thinking it worked and making it better. If you want more details on that you should contact our lab. It's really still pretty preliminary.

ADAMS: I would like to make a request. This past year I phoned several laboratories and asked them to forward to me a list of their oddball suppliers and places where you get oddball things and services so I might compile them. We are talking about very unusual things that would be useful to most laboratories.

What got me started last year is for three years I had been looking for an LED to replace those little T-1-3/4 bayonet bulbs in the microswitches that control the Faraday cups and beam steerers and finally I found a supplier. Those little incandescent lamps were running about 55, 60, 70 cents, and if you buy the premium ones it's about 80 some cents. But it's all mechanical. It's the filament that really breaks.

So the LED type is a direct replacement. The company is Lamp Technology, Incorporated, and it's 141-A Central Avenue, Farmingdale, New York 11735. The phone number is (516) 454-6464. They cost about $3 apiece, but they'll probably outlast the incandescent type by years.

BLANCHARD: Is the brightness adequate?

ADAMS: It works beautifully. I use the yellow for white, and they have green LED and red LED. This is actually one of them right here. This is the yellow LED, the bayonet type. You can see them through the little labels on the microswitches very clearly.

BLANCHARD: Good. I've been looking for this for some time, and I have never found another LED to give enough output.

LARSEN: Charlie, is that a 24-volt application?

ADAMS: This is a 28. You can get them 24, you can get them from six volts on up, in steps.

LINGREN: We have had a lot of experience with these lamps. We bought perhaps 500 of them about two years ago, and I think we paid about $1.60 apiece for them.

Some of them lasted very well, and I believe they are extremely good lamps if you power them at less than rated voltage.

We have had a lot of failures, and on checking we found we were running about 30V A/C on them and then they blackened up and died in a short while. We merely put half voltage on them, and they light with plenty of brightness.

D/C they don't last well. We bought 24 volt ones, and we found that we have been running 27 volts on them, and they don't last. But if you always drop the voltage to less than 20, they are very good.

When you first transfer to this type of a light from the incandescents you may have a bit of trouble, because the people who wired the sockets did not always put the positive D/C lead to the center conductor for the lamp. So, of course, the LEDs won't light. So then you will have to reverse the wires.

I think on the whole with our experience they last a good deal longer than the appropriate number of incandescents you could have bought for the same price.

LARSON: Some of those lamp fixtures for the microswitches have four bulbs running, and I recall melting down the plastic components under those conditions. I presume the heat output is much less with those.

ADAMS: Nothing to it. You take out the little blue covers and red covers. You don't need those any more.

LINGREN: You don't want to put your fingers on the LEDs. They are very hot.

ADAMS: Again I would ask any laboratory that has a list of things that are difficult to find, Chris sent me a list and was very helpful, Ken Chapman sent me a list, and there are one or two others whom I've forgotten that gave me some information, and I have compiled some of this together, but I'd like to make it fairly complete.

NORTON: Back to the negative ion beams, we have compiled a very informal list of the negative ion beams we found in literature that we have produced and that Argonne has produced. I brought copies of that list with me. See appendix #.

If it's convenient, I would like for people to correct the list, send it back to us, and then we will bring it to SNEAP every year or mail it off. This is any negative ion source.

LINGREN: I think it's very nice to get these discussions back to ion sources.

I mentioned to several people something that we have noticed with the cesium boiler on 860 sources, and I guess this would also hold for any other source using a cesium boiler, that it was very easy to poison the cesium especially after running lead sulfide.

When we put lead sulfide pressed into a target holder we noticed that the boiler initially works normally, but then soon stops evaporating cesium into the source.

On opening up the boiler one notices that the cesium is liquid, and our experience has been if you see cesium liquid at room temperature you should throw it away and put in a new charge to resume normal operation. I would like to hear if anyone has noticed this.

Several individuals at the meeting, were surprised to hear this.

Another point, we have found of a much more long lasting sulfur material, rather than the lead sulfide powder, which has a very short lifetime at high currents, we find if we use pyrite, which is iron sulfide, or galena, which is lead sulfide, they last indefinitely. We have pulsed them with 100 microamp pulses, two per second, for several days continuous and have not seen any erosion of these materials.

CHAPMAN: Do you assume cesium is forming a cesium lead eutectic.

LINGREN: We assume it's making an alloy of some kind.

ADAMS: You say you can get a watery type film on the cesium or it is black?

LINGREN: It's a nice, shiny looking cesium, and you wiggle the can, and it's liquid at room temperature. We know cesium is not liquid at 70 degrees but at 85 degrees.

TIPPING: We have also experienced a problem with our cesium boiler, and primarily we have been running a fluorine beam and an oxygen beam out of the source, using calcium fluoride and aluminum oxides as the materials packed into the cathodes. We have been running this source for approximately a year under the same circumstances and having no problems at all. Suddenly we have been plagued by what you just described, pouring out the cesium and having it liquid sitting on the table top. It's a new problem with us. Suggestions are welcome.

MAN IN AUDIENCE: Have it analyzed.

ADAMS: You say your packing has a binder with the aluminum oxide?

TIPPING: We are mixing it with a copper powder?

NORTON: What current do you obtain?

TIPPING: Normally we can easily inject eight to 10 microamps at the low energy end of our tandem. As far as going for maximum, I think we have had 30 or 40 microamps of fluorine.

NORTON: We got about the same amount of fluorine using lithium fluoride, no moisture.

KORSCHINEK: With such a similar high current source we are getting about two to three hundred microamps of fluorine with calcium fluoride.

SOUTHON: I would like to ask, when you pull the source apart do you clean the ionizer? What I'm getting at is the problem with the cesium, or have you also poisoned the ionizer somehow?

LINGREN: We have not ever had a poisoned ionizer.

ADAMS: Run calcium and you will.

LINGREN: When we have this problem we just empty out the cesium and put a fresh load in, and we are in business.

MAN IN AUDIENCE: Can someone suggest how to make a boron beam with the duo-plasmatron source?

ADAMS: With a duo-plasmatron use BF-3, but it's dangerous as the devil.

MAN IN AUDIENCE: I was looking for something else.

ADAMS: That's the only thing you can use.

MAN IN AUDIENCE: We have been making a positive beam using a cold cathode source and BF-3. It works, sort of. We have gotten about 50 microamps positive, but for a negative boron beams we've been using B-2 minus and B-minus and obtaining close to 100 microamps with the cesium sputter source. That's the best way to do it.

STURBOIS: The list you are putting together. Are you going to send that out to everybody?

ADAMS: Yes, I'd like to get it all together, and maybe I can get it into the proceedings.

I would appreciate anyone supplying me with what you have collected over the years and I will edit it.

MILLS: What's the secret for making a microamp calcium analyzed?

ADAMS: Calcium hydride, and it has to be made fresh. Just make it and install it in the cathode and run it.

If you get calcium hydride in the jug and keep it in a dry box, as soon as you open it, even in the dry box, you'll feel the jar getting hot, and it just doesn't last. You have to move very quickly with it.

ZIEGLER: Is the SNEAP bulletin board still active at Argonne?

MAN IN AUDIENCE: I talked to Pat at Argonne last week, and he discontinued it. It was a lot of work on his part.

BERNERS: This might be a good time for me to make an announcement. If anyone here would like to have a copy of the 1986 proceedings of SNEAP at Notre Dame, it would be a good idea to give

me your name and address at this meeting. I don't know how many
books we have that are not already assigned to someone, I think
possible around 30, at a $12 price. Once those are gone, the price
goes up to $56.

LINGREN: Would you please repeat the recipe for the lithium you
mentioned.

ADAMS: I will give you a little detail on that. Roy Middleton
is rather jumpy with noises occurring behind him. We were all in the
room, chopped up the lithium, put it on the block, lit the Bernz-o-
matic, we turned the light out in the room, and Jeff Klein had a
paper bag behind his back, and when he hit that lithium it went up in
a flame, hit the paper bag, and Roy went up through the ceiling.

MAN IN AUDIENCE: Who did he land on?

MAN IN AUDIENCE: This is on the lithium beam again. We were
running a fluorine beam, lithium fluorine, and after running that for
a while, six or seven hours, we went to see if we had any lithium
beam and we had about two or three microamps on a cathode that was
well used. I never went back to see just how this lithium fluoride
worked out.

ADAMS: Do you bake the lithium fluoride before you put it in
the cathode?

MAN IN AUDIENCE: I don't know for sure, but probably not.

ADAMS: It's hygroscopic.

MAN IN AUDIENCE: If it's hygroscopic or reactive we will do it
under hood in an Argonne bag, but however it was done it was done
very dry.

ADAMS: Is there anything more to be brought up on laboratory
reports?

STURBOIS: I was looking through the lab reports, and I noticed
a couple people having belt problems, belt problems similar to those
we had before we added a couple of screens to our charging collecting
system. Those screens are outlined in our lab report.

We have not seen any problems like that until we had two and a
half times as many hours as we ever had run on a belt. It also made
us believe it eliminated the necessity for doing a lot of belt
tensioning that had to be done as in the past.

In the past we would start to get terminal instability and would
tension the belt. That helped a little bit. Pretty soon we would be
in tensioning the belt again and installing new charge screens.

After a little while the belt would come apart or would lead to premature failure. With the new screen system we haven't had to do that. We ran 3,500 hours without tensioning the belt at all.

SESSION IX

REPORT ON SAO PAULO ACCELERATOR - E. PESSOA

(Transcribed from the tape of this session)

For the past four or five years the accelerator at the University of Sao Paulo has been working quite well. However, prior to this time we had an awful lot of difficulty but we also had the very good luck of having Mike Steal with us for eight years. By the time he returned to the States, he put us very well on the right path, so I would like to take this opportunity to thank him for that.

Since the accelerator was put in good order and I will give you some of the parameters that we consider to be good order, we did have very little utility out of it and the reason for that was the obsoleteness of our IBM 360 computer which got to the stage that it was turning the accelerator off at least three quarters of a month. The IBM technicians in Brazil that would come to fix it were such young kids, that they marvelled at the old model design that they found in the computer that we had and they really couldn't fix it. They weren't trained for it. However, this was substituted by a VAX model 780 a year and a half ago and since the VAX has been working, the accelerator has been on twenty four hours a day, twenty five days out of every month and spirits are riding very high, the data taking is really getting quite efficient.

Just to describe what we have there. There are three ion sources, the two most frequently used are dual plasmatron and a new Snick. We run at about ninety (90) KEV injection voltage. The terminal usually, almost exclusively, is run at eight (8) million volts. Sometimes if someone is patient enough, they can take it up to eight and a half. The energy stabilization is less than a killovolt for protons. The transmission is very high. Professor Sala is extremely pleased with it. It is running at ninety (90) percent for protons and it drops to seventy five (75) percent for oxygen ions. The chains have been used for about thirty five thousand (35000) hours and with no apparent stress. The tube itself is a little over forty thousand (40000) hours and once again no apparent stress. It is rather eutopian for us, although I must admit, we deserve it after all the past troubles at least this year.

However, this rather remarkable situation does have a little bit of shadow in the future. For example, our supply of the sulphar hexafloride is low. The import/export difficulties that prevail I am sure in most Latin/American countries. We don't forsee the gas arriving too soon and the machine engineer told me that quite soon he is going to insist that no limit over seven (7) million volts on the terminal be passed. Most of our problems do stem from the isolation, the economic difficulties, the availability of new technological developments and components. I mean, a situation I am sure that is familiar to most of you.

However, there is a new program which Professor Sala has been thinking about, since the accelerator is running so well now. He has been thinking seriously in the past few years about upgrading energy and that we ought to do something and that we shouldn't just stagnate here. After serious thinking, he decided that perhaps the best thing would be a booster of the super conducting linac type. His choice and that of the people that he consulted, the choice was made on several factors, one being for this type of machine, we do have the available space. The second thing is that the Brazilian National Company of Metals and Mines which has alot of influence in obtaining federal funds for projects like this is more than happy to support any sort of a project that involves niobium and this is a very important asset in obtaining funds in Brazil and for this reason, his attention or interest was immediately steered in the direction of the niobium split ring. With that, the third thing that is very important in Brazil, he needed some sort of project that has been tested and proven as essential. We can do no technological developments, it would be impossible. The idea will be to copy letter for letter perhaps if there has been some updating in the past few years, but the idea as I understand it to copy letter for letter the Atlas project from the Argonne Laboratory. The funds for this I think our out but I must admit that people do not like to discuss the situation of their funds and so for discreetness, I can say nothing here but I do believe things are okay.

Finally, as far as the Linac project is concerned, what we have done so far and this is all in cooperation with the Argonne National Laboratory, they have sent their designs and they have been extremely nice to us, we built the pre-buncher, the one gap harmonic pre-buncher, that has been built, the supporting table is practically finished for ours is a vertical machine of course and their's is a horizontal machine. It should be installed within the next few months if everything goes according to schedule. We plan to put from the system optics the calculations we have done, in fact that is something I would like to discuss while I am here. We plan to put it one point two meters (1.2) in front of the first accelerating electrode, upstream. There is a little question, the voltage that will be necessary if we want twelve (12) megahertz bunching which Professor Sala would like, so that the Pellatron could continue with the time of flight program. But, that is something I would like to see while I am here.

The architects for the building have been met with several times. There is a company, Oberex, who did the buildings for the reactors at Ungerdar, the nuclear reactors and so they are a company that are familiar with building technological buildings and though this is a very small project for them really, they are quite interested in doing it so I really think that it should go quite smoothly because we do have someone with superb experience in that type of thing. The money for the building has come through. It is the University of Sao Paulo that will sponser the building money and that presumably will be started if all goes according to schedule, January/February the construction will start.

I think that that more or less summarizes the present state of the Pellatron Laboratory and its future plans at the University of Sao Paulo.

Discussion Following E. Pessoa Paper

ROWTON: Questions?

CHAPMAN: If I understand you correctly, you have the facilities fairly easily available in Sao Paulo for the explosive bonding of niobium and copper that would be necessary for this, is that correct?

PESSOA: I think the Niobium Division of the Brazilian company of Metals and Mines does, yes.

LARSON: I can't help but be struck by your comments to the effect that you couldn't do any development work there. I can appreciate what you are saying, and I think the plans are very good.

But this is a matter of simply plunging in and doing something, whether it is on the forefront or the back burner. I think mental attitude, among the people involved is very crucial.

PESSOA: Well, maybe I was a little strong in that. It isn't that we do nothing unless it is sent to us. If anything goes wrong, we'll do our best to put in innovations and try to fix it up, but the point is we cannot be involved, in long-term development projects. We don't have the people. We don't have the raw materials. We don't have the equipment. We don't have the components. We are isolated.

What would take, say, a year here would take 10 years there. It could be done, but it would take 10 years.

Now once this project gets along, very slowly, of course, it's something that develops naturally. You have to start and sooner or later do some sort of development, but it has to be done in a natural way. It can't be part of the plans or we would be unable to finish it in any reasonable time. I believe the plans are that in five years Professor Sala would like the superbuncher and three section of six resonators each completed and working. Maybe that's too ambitious.

ZINKANN: Are you planning to fabricate resonators there?

PESSOA: Eventually, yes. But certainly not the first few. Naturally, they have niobium of very good purity.

ZINKANN: They do electrode beam welding?

PESSOA: That I don't know. I'm sorry. I don't know enough about it yet. I just arrived to spend three weeks at the Linac laboratory here to learn myself what's going on.

NEEDHAM: Are you going to use both the 48 and 12 megahertz bunching?

PESSOA: That's the hope, yes.

NEEDHAM: Have you built the buncher and has it been tuned?

PESSOA: It hasn't been taken out of the machine shop and mounted yet, no. Why? Can you foresee any trouble?

NEEDHAM: Well, I have one exactly like it sitting on my desk at home that I am about to tune. I am curious as to how easily the tuning went. I know about some travails that go on at Argonne in that process.

PESSOA: I wish I could help you. Perhaps later you could give me the names of some people at Argonne that we could talk to about that.

PARDO: I can tell both of you, since they are room temperature devices there will be almost no trouble.

WEIL: Yes. You spoke of running short of SF_6. What's the supply of helium like now? Do you have to bring that in?

PESSOA: Well, the supply of the helium gas I think will be all right, because we have White Martine's company there, and certain companies that presumable will be able to facilitate for us the importation.

So I think the helium will not be as troublesome as the SF_6, which is gratifying. It will come. It always does.

REJUVENATION OF MONOCAST
NYLON LADDERTRON LINKS

R. Bundy, R. Lefferts, J.W. Noe,
J. Preble, and A. Sullivan

Physics Department, SUNY at Stony Brook
Stony Brook, New York 11794-3800

SUMMARY

Following a Laddertron failure, difficulties with the
chain re-assembly led to a new technique for reforming
the elongated bushing holes in the nylon links to the
original dimensions. This report will review the history of the system and discuss the novel technique.

INTRODUCTION

On December 18, 1986 we had a total failure of our
Laddertron charging chain just after it had come up to full
speed. Gradually increasing vibration readings had been
apparent for a few weeks before this, culminating in a trip
of the vibration interlock on the 18th. While investigating
the cause of the trip the Laddertron broke.

Fortunately the Laddertron break caused only minor
damage in the tandem: the terminal and base inductor supports broke off and a thin aluminum shield adjacent to the
mid-section idler pulley was chewed up. The inductor supports are intended to give way easily to protect the inductors, and indeed the inductors were only slightly scored.
Although it is well known that NEC Pelletron breaks generally cause only minor damage, with the HVEC Laddertron we
initially feared much worse because of its much greater
weight (3.5 pounds/linear foot).

SERVICE HISTORY

Our experience with the Laddertron is summarized in Table 1. Rebuilds have been done after about every 4,000 operating hours, on average. There was one previous total failure, at 13,900 hours, also with no serious damage to the tandem. Not shown in the table, several times single links in the chain broke without causing a complete failure. The possibility for such a graceful failure is of course a major advantage of the dual-chain approach.

Table 1.   Service History of the Stony Brook Laddertron.
==

Date	Hours	Work Performed ....
6/80	[66,300]	Acceptance tests (250 uA @ 10 MV).
1/81	1,500	Service drive and terminal bearings.
7/81	2,475	Rebuild (#1) with new bushings.
4/84	10,185	Rebuild (#2) with new bushings and replace all pulley bearings.
7/85	13,914	Rebuild (#3) after major break (rotate link bushings 180°).
4/86	16,840	Service drive and terminal bearings.
9/86	19,126	Rebuild (#4) with rotated bushings.
1/87	20,153	Rebuild (#5) with new bushings after major break. Boil links to renew.
2/87	20,800	Find one link pin missing.
6/87	22,350	Resurface terminal pulley.

==

Low-frequency vibration of the base pulley is continuously monitored by an accelerometer[1] and typically the need for a rebuild can be forseen weeks or months in advance by gradually increasing vibration readings. Sometimes this is also associated with the appearance of a 1 Hz "signature" in the terminal ripple and/or the charging current. A more serious warning sign is a drop in the running tension from the normal 850 pounds[2] to 800 pounds or less due to impending link or bushing failure. (These numbers reflect a few hours of warm-up; readings in the first few minutes are up to 100 pounds higher.) Normally we find that after a few days of pronounced stretching, the warm length (and hence the tension) of a new or rebuilt chain will be stable for six months or more.

THE LADDERTRON REBUILD

We proceeded with our normal Laddertron rebuild, which we are now typically able to complete in about 15 person-days. This goes roughly as follows:

All parts are first disassembled and washed in mild soap and warm water. The Laddertron chain has about 300 units, each of which consists of two s/s cast rung ends connected by a machined aluminum rung, and two monocast nylon dogbone links. Including also the 4 screws, 4 stainless-steel bearing pins and 4 lead-loaded teflon bushings[3] there are altogether 21 pieces per unit, for a total of 6,174 distinct pieces.

The link bearing pins are buffed on a wheel using a levigated alumina polishing compound, and any noticeably marred pins are discarded. The bushings are a standard metric type whose active component is lead-loaded teflon impregnated into phosphor bronze.[3] In several of the rebuilds we were successful in simply rotating the used bushings 180° to get a fresh surface, but usually they have just been replaced outright.

For the re-assembly, sections of seven rung ends
(not complete rung units) are stretched in a hydraulic
jig to a standard tension of 500 pounds or so, and
their lengths are recorded. Segments of the same
length (equal within 5 mils) are then paired for the
final assembly, which is also done under comparable
tension. The re-assembly jig is a copy of one loaned by
HVEC for our first rebuild.

THE BOILING-WATER TREATMENT

During the re-assembly this time we first noticed that
the length variations of the seven piece sections were much
larger than usual. This lead to a closer inspection of the
parts, and revealed considerable stretching of the holes in
the nylon links. The elongation was typically 8 mils, but
in some cases as much as 13 mils (0.325 mm), with almost no
links retaining their original roundness.

On a hunch, to see if this might affect the shape, we
decided to try boiling the links in water for a half hour
or so to re-hydrate the nylon. (This idea came in part from
the observation that model-airplane hobbyists sometimes
boil plastic propellers to restore their strength.) Remarkably, this simple procedure seemed to have a dramatic re-
generative effect on the nylons.

Before proceeding to boil our entire set of links we
consulted with the DuPont company on the properties of
monocast nylon. The problem seemed to be one of having
allowed the nylon to dry out. If first dried out and then
subjected to stress nylon will deform and maintain its new
shape. Dupont agreed that it was possible in principle to
restore the nylon to its original condition by boiling, but
they were as surprised as we were that the "memory" of the
nylon could be as long as seven years, especially in an
environment with less than 5 ppm of moisture.

STATUS OF THE ATLAS POSITIVE-ION INJECTOR PROJECT

R.C. Pardo, R. Benaroya, P.J. Billquist,

L.M. Bollinger, B.E. Clifft, P.K. Den Hartog,

P.J. Markovich, J.N. Nixon, K.W. Shepard, and G.P. Zinkann

Argonne National Laboratory, Argonne, IL   60439

ABSTRACT

The goal of the Argonne Positive Ion Injector project is to replace the ATLAS tandem injector with a facility which will increase the beam currents presently available by a factor of 100 and to make available at ATLAS essentially all beams including uranium. The beam quality expected from the facility will be at least as good as that of the tandem based ATLAS. The project combines two relatively new technologies - the electron cyclotron resonance ion source, which provides high charge state ions at microampere currents, and RF superconductivity which has been shown to be capable of generating accelerating fields as high as 10 MV/m, resulting in an essentially new method of acceleration for low-energy heavy ions.

INTRODUCTION

The existing ATLAS facility provides beams of heavy-ions through approximately mass 130. Available energies range from over 20 MeV/A for lighter ions down to approximately 5 MeV/A for mass 130. In discussions with our user group concerning future program needs, two major areas of focus emerged. The first was a desire to increase the beam intensity for all ion species available by a factor of ten to one hundred over what is possible from our present negative-ion source and tandem injector. The second was to obtain beams of 7 to 10 MeV/A energy for all possible masses through uranium. These features were desired without compromising the present qualities of the ATLAS facility: good beam quality, ease of operation, and continuous wave (cw) operation.

The Argonne Positive Ion Injector (PII) project[1,2] will address these goals by replacing the negative-ion injector and FN tandem with an electron cyclotron resonance (ECR) positive-ion source and a superconducting linac of a new design which makes use of the high field gradients possible from superconducting structures. The total effective accelerating voltage for the completed Positive Ion Injector will be approximately 12MV. The beams from this new injector accelerator will then be further accelerated in the existing ATLAS superconducting linac. A schematic representation of the completed facility is shown in Figure 1. Figure 2 is an enlargement showing the Positive Ion Injector in more detail.

The positive ion injector project will progress in three phases. In the first phase, which is underway, the ECR source system, a 3MV prototype linac, and a building addition to the present accelerator area is being constructed. Even this small injector will provide sufficient energies and beam currents to exceed the tandem performance in some ways. Phase II will increase the total voltage of the PII to 8MV. Finally, the complete uranium capability of the machine will be realized in Phase III with the enlargement of the PII to a total voltage of 12MV.

The activities required to complete Phase I of the project include most of the developmental goals required in the project. The ECR source system is nearly operational, including the high voltage platform. The prototypes of all new resonator designs are well along with three resonators tested at this time. The building addition design is complete. Construction will begin in the Fall, 1987 and occupancy is expected in early spring of 1988. The first test beams from the Phase I PII project is anticipated at the end of 1988 or very early in 1989. In the remainder of this report, a more detailed description of the project status will be presented.

THE ECR POSITIVE ION SOURCE

The electron cyclotron resonance (ECR) positive-ion source has been demonstrated to be an exciting new technological development in the production of extremely high currents of high charge-state heavy ions. Much of the early ECR development has come from the efforts of Geller[3]. The beam properties of the ECR source which make it an attractive option for this application are:

1) high charge state heavy ions,
2) small energy spread (apparently less than 5Q ev),
3) good transverse emittance ($\gamma\beta\epsilon = 0.2\text{-}0.6\pi$ mm-mr),
4) high operational reliability, and
5) large beam currents in comparison to a negative-ion source.

For the PII project, the ECR positive-ion source is mounted on a high-voltage platform providing up to 350-kv potential for pre-acceleration of the ions. This will produce, for example, $^{238}U^{20+}$ of 7 MeV with a velocity of .008c. The ions will be bunched in a two stage bunching system providing a pulsed beam with a time width of the order of 0.3ns for injection into the linac.

The design was driven by a desire to have a source with the good charge state and current characteristics of the large ECR sources presently operating and with the additional requirements of:

1. low total power consumption,
2. production of ions from solid materials,
3. good transverse emittance,
4. low total energy spread, and
5. low material consumption modes of operation.

The relevant source parameters are listed in Table 1. The detailed issues of the design of the PIIECR has been discussed elsewhere[4].

In Figure 3, the charge state of ions expected from an ECR source across the periodic table are plotted on three curves representing three different levels of electrical beam current. The curves shown represent a smoothed, by eye, fit of published information for DC operation of heavy-ion ECR sources as of early 1986.

A most important requirement for our application of ECR source technology is the need to obtain ions from solid materials. ECR sources have historically operated only with gases. In the last three years, a number of laboratories have begun to use solid material feeds to the sources. The experience from these laboratories is quite encouraging and for many solid materials which have been attempted, one sees that the results fit quite nicely on the curve of Figure 3. Use of solid material as the primary feed material for the ECR source is an extremely high priority for this project. Therefore, a significant effort has been expended in order to make the source design as flexible as possible for accepting solids.

A photograph of the system as it is being constructed now is shown in Figure 4. The source, platform, and analyzing beamline on the platform are nearly complete. We are presently awaiting the final installation of the RF transmitter. Magnetic field maps of the solenoid fields, hexapole fields, and the 90 degree analyzing magnet have been made. The hexapole fields are weaker than had been anticipated due to an error in the procurement specifications. This error may be corrected later in the project. Early test operation of the transmitter uncovered a number of problems with the delivered system. First operation of the source will occur in October, 1987.

The beams of high charge-state ions, with energies up to 350*Q keV, expected from the source system may be unique in the world. These beams have generated a significant demand for their use in the atomic physics group at Argonne and elsewhere. A temporary beamline has been constructed in the room housing the ECR source system. This beamline will be used to perform experiments in atomic physics with beams from the ECR source system during the next year. The beamline will be moved next summer into the extension to the accelerator hall which is now being constructed.

THE LINEAR ACCELERATOR

The linac design adopted is a superconducting linac of independently-phased resonators. The first stages of acceleration will consist of short resonator structures followed by superconducting solenoids to refocus the beam. The initial cell length can be short because of the use of superconductivity for both the RF resonators and transverse focusing solenoids. The choice of radially focusing solenoids and the separation of these focusing elements from the acceleration region is quite important in minimizing the coupling between the radial and longitudinal phase space.

In Phase I, the first of eventually three cryostats will be constructed and installed. Cryostat one will hold five resonators and three solenoids and will provide a total of 3MV of accelerating voltage. A picture of the low velocity region of the positive-ion injector for the Phase I PII is shown in Figure 2. The cryostat design is nearly complete and construction is expected to begin this Fall. Delivery of the first cryostat is anticipated in March, 1988.

A total of four types of resonators will be necessary in order to efficiently couple to the beam as the particles rapidly gain velocity.

These types are characterized by the velocity at which the particles experience maximum accelerating voltage. For the PII linac resonators with matched velocities of β= 0.009, 0.015, 0.025, and 0.037 are needed. The sequence of resonant cavities shown in Figure 2 for Phase I will be modified as the PII is expanded in Phase II and Phase III. Additional resonators of the β=0.015, β=0.028, and β= 0.037 classes will be required for acceleration of the heaviest ions.

LOW-VELOCITY ACCELERATING STRUCTURES

The requirements for accelerating low velocity ions constrain the possible resonator designs[1]. The coaxial quarter-wave superconducting resonator has been selected[5] as the resonant structure. Each cavity will have four accelerating gaps formed from three drift tubes. In order to minimize the bunching requirements it is desirable for the resonator frequency to be as low as practicable. The low resonant frequency is also necessary in order that the transit time factor for these low- velocity (.008c) particles be large but at the same time maintain as long an accelerating structure as possible. Therefore the design frequency of 48.5 Mhz, one half the 97 Mhz frequency for the ATLAS low-beta (0.06c) and high-beta (0.105c) split-ring structures, has been chosen for these structures. The superconducting material is niobium and the fabrication techniques rely heavily on the experience of the ATLAS project.

Construction of the lowest matched velocity unit was completed in October, 1985 and the unit has performed stably at fields of up to 10 MV/m. The matched velocity for the first cavity is .009c and is the only resonator of this type which will be needed, even for uranium acceleration. The vibrational levels measured for this resonator are low enough that phase control with our present techniques can work well. Based on these tests, we are now assuming that this resonator will be able to operate online at accelerating fields of 4.5 MV/m.

The next two types of accelerating structures will have a matched velocity of 0.015 and 0.025. The prototypes of these two classes of resonator are complete and were tested at fields up to 6 MV/m. Our accelerator designs assume these resonators will operate at 3 MV/m.

The last class of resonator needed will have a matched velocity of $\beta=0.037$. The prototype of this class is under construction. Test are expected in early spring of 1988. The entire family of structures planned for this lowest velocity section is shown in the magnification in Figure 2.

PERFORMANCE

The expected energy performance of the Positive Ion Injector is shown in Figure 5. The three phases of the project are compared to that of our present 8.5MV tandem. Figure 6 shows the energy performance of ATLAS during the various phases of this project while Figure 7 emphasizes the beam current capabilities resulting from the project at a representative energy. These results depend strongly on the actual performance of the ECR source and assumes that the $\beta=.009$ resonator will operate at 4.5 MV/m and all other structures will operate with field strengths of 3.0 MV/m.

These performance projections assume charge states and currents taken from Figure 3. For heavier beams, one to two additional stripping foils are also assumed to be used in these calculations. The use of foil stripping and higher charge states from the ECR source are the cause for the loss in beam intensity of heavier ions that is indicated in Figure 7 for the earlier phases of the project. It is apparent from figures 6 and 7, especially, that even the small 3MV injector prototype of Phase I allows a new regime of current and energy to be accessed that is unavailable with our tandem.

Acknowledgment

This research was supported by the U.S. Department of Energy, Nuclear Physics Division under contract W-31-109-ENG-38.

References

1. Bollinger, L. and Shepard, K.W., Proc. of the 1984 Linear Accel. Conf., Seeheim, Fed. Rep. Germany, May 7-11, 1984, 217 (1984).

2. Pardo, R.C. Bollinger, L.M. and Shepard, K.W., Nucl. Inst. and Math. in Phys. Res., B24/25, 746 (1987).

3. Geller, R., I.E.E.E. Trans. on Nucl. Sci., NS-23, 904 (1976).

4. Pardo, R., et al., Proc. of the VII Workshop on ECR Ion Sources, Julich, Fed. Rep. Germany., May 22-23, 1986, 223 (1986).

5. Shepard, K.W., Proc. 1986 Linear Accelerator Conference, SLAC-303, Stanford, California, June 2-6, 1986, 269 (1986).

Table I

Argonne PIIECR Ion Source Parameters

Magnetic Field

Peak on Axis	4.75 kG
Solenoid Magnet Power	35 kW
Maximum Solenoid Current	500 Amp
Typical Solenoid Current	360 Amp
Mirror Ratio	1.60
Mirror Ratio Range	+/- 0.2
Length of Second Stage Mirror	47 cm.
Hexapole Material	Nd-Fe-B
Number of Poles	12
Hexapole field at chamber	3.8 kG

RF System

Frequency	10.5 GHz
RF Power	2.5 kW
Independent Control	both stages

Dimensions

Solenoid ID	21.6 cm.
Solenoid OD	64.8 cm.
Hexapole ID	12.0 cm.
Hexapole Length	49.5 cm.
Vacuum Chamber ID	10.8 cm.
Anode Aperture	8.0 mm.
Extraction Aperture	10.0 mm.
First Stage Aperture	12.0 mm.

Figure Captions

Figure 1: Plan view of the ATLAS facility and the Positive Ion Injector project.

Figure 2: Detailed view of the Positive Ion Injector. The magnified region shows the ECR source and the family of low velocity resonators which are planned for the PII project.

Figure 3: Charge state distribution from ECR sources. Data for this figure were taken from a wide variety of published and unpublished literature. The smooth curves for three different current intensities are used in calculation of the expected performance of the PII and ATLAS.

Figure 4: A photograph of the ECR ion source on its high voltage platform.

Figure 5: Expected beam energy from the Positive Ion Injector as a function of mass for the three project phases. Termination of the tandem curve at mass 120 reflects the inability of the tandem to transmit adequately heavier beams.

Figure 6: The beam energies expected from ATLAS with different injector assumptions but with an assumed beam intensity of 3 pnA.

Figure 7: Beam current from ATLAS versus mass number for the various possible injector types at a fixed energy of 5MeV/A. The beam currents shown reflect the choice of charge states required from the source for a given total accelerating voltage and the need to perform additional stripping to higher charge states for the various configurations.

270

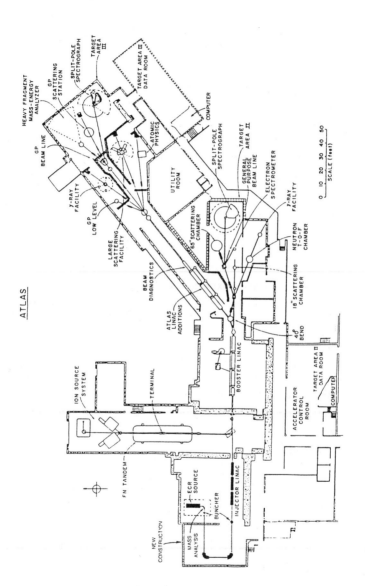

ATLAS

FIG. 1

ANL-P-I8,482

ARGONNE POSITIVE ION INJECTOR SYSTEM

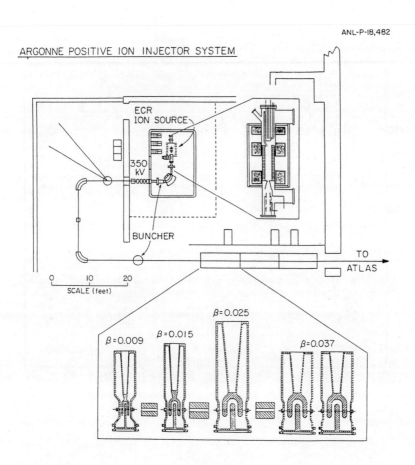

FIG. 2

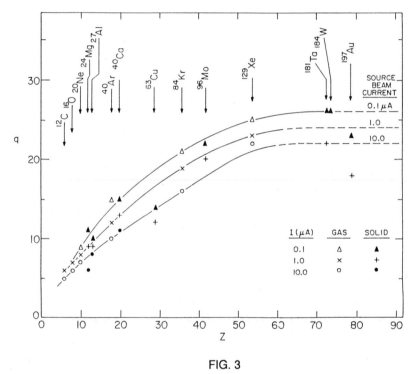

FIG. 3

FIG. 4

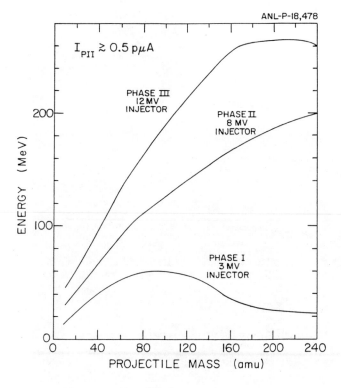

FIG. 5

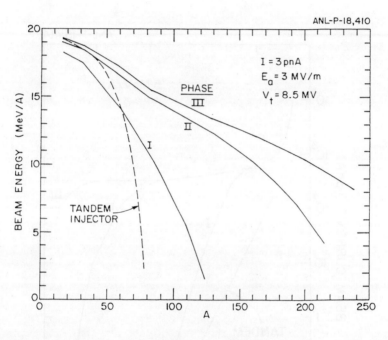

FIG. 6

276

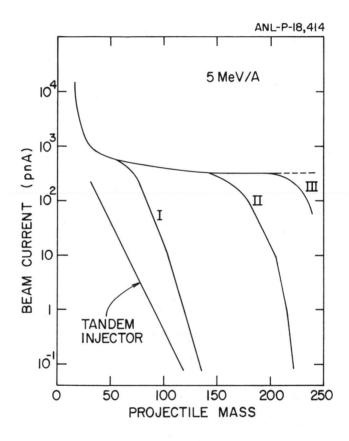

FIG. 7

Discussion Following R. Pardo Paper

STORM: Thank you very much for a very interesting talk, about in the required time. Do we have any questions?

MC KAY: Who made your isolation transformer?

PARDO: Those were made by Hipotronics. They and Universal Voltronics bid on the package, and Universal Voltronics was over priced by a factor of two to three.

Maybe I might just mention the tests we have done up to now. Transformers have been working and providing power quite nicely, with no problem. This is without high voltage, in all cases. When we did the high voltage tests there was essentially no current load on the system at that point. It was too early into the game.

We had the Glassman supply, and we wanted to make sure it actually worked, and so we hooked it up and tested it, we had a small 120 kilovolt voltage divider network and placed that on it and looked at ripple.

At 120 kilovolts we saw less than one-half volt, and in fact we were pretty convinced most of what we were seeing was our own pickup, not really doing a very good job of measuring what Glassman was really putting out.

Then we hooked it up to the platform and saw I think it was like a volt or volt and a half ripple with the high voltage transformers wired up, that is the distribution lines had been hooked up to them and so forth, but essentially drawing no power.

At 350 kilovolts total leakage current, and I think it had to be essentially all through those transformers, was I believe about 200 microamps of leakage. No symptoms of corona, microdischarges or anything. We couldn't tell the supplies were on except by looking for a volunteer.

Actually we laughingly convinced ourselves because both Pete Belquist and John Bogardy got rather severely shocked later on by that little 120 kilovolt divider that they had disconnected and tried to pick up, there were some compacitance that had charged up, so we know we put charge on it. That was how quiet things were, and that was at 61 per cent relative humidity.

One of my claims had been that this is a positive voltage, that we would have much less corona problems with the positive voltage system than with the negative voltage system, and at least the first tests certainly indicated.

HYDER: What would you do to get power on the solenoids down from those?

PARDO: Geller uses no iron, so the first and dominating step is to put iron in, and the next step is just to do a careful magnet design. Geller used coils from an old Plasmamerra machine that has very poorly designed coils.

I had a colleague, Ken Thompson, over in High Energy Physics Group, who had spent 20 years building magnets of various types, crank up his programs and go through minimization on conductor size and all that sort of stuff. So we gained a little that way, but mostly it's just putting iron in.

LETOURNEL: What was the total power?

PARDO: It looks to me as though we will run the source as designed for under 60 kilowatts of total power, and that includes all the analyzing magnets, the RF transmitter, the solenoidal coils, and all the ancillary turbo pumps and everything.

The solenoidal coils themselves came in exactly as designed, 31 kilowatts, at design fields of actual power.

KORSCHINEK: Do you intend to use this source for accelerator mass spectrometer measurements?

PARDO: We intend to try.

KORSCHINEK: According to the lab surveys you expect some hot and cold areas, according to the magnetic fields...

PARDO: Yes.

KORSCHINEK: ...so the first question is what do you think about memory effects.

That's the first question. The other question is you speak about conversion efficiency for the ions. I took just your numbers you mentioned. If you do it very well, maybe you are in a range of like one per cent. Is that right?

PARDO: Yes. In fact, one percent is about right for the current crop of things. People have built some smaller ECR sources, very specialized, like a lithium source, and I think that was at Karlsruhe, and there they have achieved conversion efficiencies of 30 to 50 per cent.

So it's possible due to some very careful designing to get some very good results.

The memory effect is a real problem. No question about it. The large surface areas are potentially a problem. It might be a little less of a problem than you would at first realize, in that the radio confining fields of the hexapole force the ions to very specific places. The patterns will be triangular, sharp triangular cusp-like shapes, and they resputter the material or re-evaporate the material, is probably a more appropriate phrase, off the surfaces, but that produces a memory effect.

That's partly how one keeps the efficiency as high as it is for a volume as large as this.

So the source has surprisingly good efficiency for such a monster in physical size, and yet the memory problem is certainly a real problem.

When I built the source I built it with the thought that we should be able to put liners inside the source, but as you can imagine even with the liner in the source I think a turnaround time for the source on the order of two to four hours, for example, as a minimum that one would expect in that sort of situation, at least as I envision it right now.

So we hope that we can learn a number of things about how to make ions in a source of this type.

For accelerator mass spectroscopy I think an interesting question will be to look very hard at the first stage. Right now people have done the first stage from a plasma perspective, but if you view the source as I described it in passing as an ion source plus a stripper you can begin to imagine whether you can do something different to the first stage, whether you can make solid materials in the first stage even if you maybe don't have little beam currents that you might often get. For AMS it still might be an interesting way to do it. You have now a much smaller region to worry about in exchanging and in contamination.

In other words, the assumption I'm making is that the contamination comes mostly from neutral particles, and not from the ionization particles. Once they are ionized they should be pretty well confined in the magnetic fields.

I'm assuming that contamination and efficiencies are effects of us not being able to ionize the vast numbers of neutrals that are put into the system at that point.

So these are the kinds of things we want to begin to study. I would not claim we have made even a serious conceptual development on what to do, but the general idea for me is to look very hard at that first stage and try to see if there aren't some novel ideas that up to now have not done, ways of generating solid material ions in that first-stage region, and being willing probably to take lower beam currents under certain conditions just to get the better efficiencies and easier exchange of beams and so forth.

STORM: Any other questions? I would like to congratulate you. The first I heard about this project was about two years ago, I think, and I don't think you had done very much at all at that point.

PARDO: That's right.

STORM: It seems you have accomplished an incredible amount in a two-year period. I hope you are able to keep up something like the same velocity.

THE FLORIDA STATE UNIVERSITY SUPERCONDUCTING LINAC

E.G. Myers, J.D. Fox, A.D. Frawley, P. Allen,
J. Faragasso, D. Smith, and L. Wright

Florida State University

As early as the fall of 1977 it was decided that the future research needs of our nuclear structure laboratory required an increase in energy capability to at least 8 MeV per nucleon for the lighter ions, and that these needs could be met by the installation of a 17 MV tandem Van de Graaff accelerator. The chief problem with this proposal was the high cost - at least $11M in 1977 dollars - and the lump sum nature of the funding involved.

It became apparent that a far less expensive option was to construct a linear accelerator to boost the energy from our existing 9 MV tandem. The options open to us among linac boosters were well represented by the room temperature linac at Heidelberg and the superconducting Stony Brook and Argonne systems. By the Spring of 1979 it had been decided that both capital cost and electric power requirements favored a superconducting system. As regards the two superconducting resonator technologies - the Argonne niobium-copper or the Caltech-Stony Brook lead plated copper - the Argonne resonators, though more expensive to construct, had the advantages of more boost per resonator, greater durability of the superconducting surface and less stringent beam bunching requirements. Additionally, at that time, sections of the Argonne test-booster, similar to our own requirements, were already in operation.

In 1980 pilot funding from the State of Florida enabled the construction of a building addition to house the linac and a new target area, and the setting up of a small, three resonator, test booster. Major funding by the NSF for the laboratory upgrade started in 1984. With these funds we purchased our present helium liquefaction and transfer system and constructed three large cryostats, each housing four Argonne beta = 0.105 resonators and two superconducting solenoids.

The last large cryostat was completed and installed on-line
early this year and the linac was dedicated on March 20. Nuclear
physics experiments using the whole linac began in early June. The
total cost of the accelerator upgrade has been approximately $2.75M,
of which approximately $0.75M was provided by the State and $2.00M by
the NSF.

OUTLINE OF THE LINAC

The layout of the accelerator in its present form is shown in
fig. 1. Ions from a sputter source operated at 120 KV are bunched at
48.5 MHz (half the linac frequency) in a gridded, single gap buncher,
based on an Argonne design[1], before injection into the FN tandem.
The tandem is equipped with the usual foil stripper in the terminal
and a second stripper midway in the high-energy acceleration tube.
Leaving the tandem the accelerated beam passes through a 24.25 MHz
chopper, which removes the residual beam between pulses. The beam
pulses are energy analysed in a 90 degree magnet and then pass
through a 48.5 MHz phase detector[2]. The output signal of the phase
detector is compared with the linac master oscillator in a double
balanced mixer. The output of the DBM is then fed back to the phase
control of the pretandem buncher so as to ensure the beam pulses
arrive at a constant phase with respect to the linac. At this point
the pulse width is approximately 1 ns. The beam is "super-bunched"
by a single beta = 0.06 resonator to form a time focus of about 100
ps at the entrance to the main accelerating section (see fig. 2).
This consists of 12 beta = 0.105 resonators. Finally the energy-time
structure of the pulses can be modified in a single beta = 0.105
rebuncher resonator. A switching magnet then steers the beam to one
of the experimenters' beam lines.

CRYOSTAT DESIGN AND RESONATOR DRIFT TUBE COOLING

All of our cryostats have been designed, and except for work
requiring large rolling and milling facilities, were constructed, in
house[3]. The essentials of our large cryostats are shown in figs. 3
and 4.

The cryostat has a "bath tub" design with all the contents
hanging from the vacuum vessel lid. The resonators are hung upside
down from the liquid helium dewar in such a way that the hollow
niobium loading arms fill with liquid helium and the copper outer
case is cooled by conduction. Effective cooling of the drift tubes
requires removal of the gas bubble which becomes trapped there, due
to the loading arms bending back on themselves. Our solution to the
problem of removing this gas bubble has been to insert into each
loading arm a plastic tube, whose end connects to the helium gas
return line, but at a point at which the pressure is less than that
of the gas inside the dewar. This pressure difference is produced by
a simple check valve, consisting of a teflon slug, which sits on top
of an orifice in the gas return line.

The required check valve weight can be found using a simple hydrostatic argument (see fig. 5). In general, in order to ensure that venting of gas from the drift tube at A is initiated, it is necessary that the check valve can support a head of liquid all the way from the bottom of the loading arm at B, to the point where the vent line joins the gas return line at C. This implies we choose $mg/a = (\rho_\ell - \rho_g)(h_1 + h_3)$. Unfortunately, when there is no gas to vent at A, liquid will flow through the vent line, driven by a pressure $(\rho_\ell - \rho_g)gh_3$. Direct return of liquid to the refrigerator increases the mass throughput it is required to handle, and leads to reduced efficiency.

As a practical compromise, we have been using a check valve which is capable of sustaining a head $(\rho_\ell - \rho_g)gh_1$, when the dewar is half full of liquid. In our large cryostats this pressure difference is 0.058 psi, and the teflon slug weighs 14 grams. We have estimated[4] that once gas flow has started, and the vent line (consisting of 31" of 1/8" I.D. tube) is full of gas, a mass flow of up to 0.4 g/s can be sustained, equivalent to an evaporative cooling rate of 7 W per drift tube.

So far our experience with this check valve weight has been less than satisfactory. We have found that drift tube venting often fails to start, limiting resonator fields to < 2 MV/m. On the other hand, depending on the sequence in which the resonator field levels are raised, it is possible to initiate drift tube venting, and so run all the resonators in one cryostat close to, or above 3 MV/m. Unfortunately, under these circumstances, there is an excessive return of liquid He from the cryostat, and very poor refrigeration performance.

We are continuing to work on optimizing the check valve system and have recently installed a heavier valve, but with a flow restrictor (at point C in fig. 5), in one of the large cryostats. As an alternative, we are also considering filling the cryostat dewar through the loading arms (see fig. 6).

CRYOGENIC SYSTEM

Our refrigeration system consists of two helium liquefiers (Koch Process Systems models 1630 and 1480) fed by three (KPS) 200 SCFM rotary screw compressors. These supply liquid helium to a 1000 liter dewar and from there to the linac via an overhead transfer line. The connections to the cryostats use demountable joints. We are able to maintain the whole linac cold using the 1630 alone, using liquid nitrogen precool a very small fraction of the time, or else with the 1480, using precool under 50% of the time. This is consistent with a total refrigeration load of 110 W. We had estimated a quiescent load of about 70 W, based on 10 W per large cryostat (standing loss measurement of 8 W, plus 2 W per demountable joint) plus a measured 40 W for the distribution line and buncher cryostat together. The

reason for this discrepancy is not known. The extra cooling required
when the resonators are excited has been supplied, so far, by
increasing the fraction of time spent on precool. It is intended
however, whenever possible, to operate both refrigerators in
parallel, giving an estimated capacity of nearly 200 W without liquid
nitrogen precool. A control system to ensure stable operation of the
two refrigerators in parallel is presently being tested. The
redundancy in our two refrigerator, three compressor system has
allowed periodic maintenance of the refrigerators and compressors to
be carried out, while keeping the linac continuously cold throughout
the whole year. The total LN_2 usage of the laboratory was 180,000
liquid gallons for the year 7/1/86-6/30/87.

RESONATOR CONTROL ELECTRONICS

The fourteen sets of resonator local control electronics have
been constructed in house and are a major achievement of our small
electronics staff. Except for certain component and layout changes,
they closely follow Argonne designs. Extensive debugging and
refinement of setting-up procedures has been time consuming.
Particular difficulty has been experienced in maintaining fast
risetimes on the Pin Diode Pulsers (which drive the Voltage-
Controlled-Reactance fast tuning devices) at high resonator field
levels.

OPERATION

As of September 87, we have accelerated beam with the buncher
and 10 of our accelerating resonators, two resonators being down due
to minor electrical problems in the cryostats. So far our
experimental program has called for beams of 95 MeV ^{29}Si (single
stripped from a natural Si cone), 70 MeV C, 96 and 110 MeV O and 50
MeV Li, each requiring 6 MV or less of linac boost. Consequently
achieving high field levels has not been given a high priority in our
initial running phase. We have been able to run, except for
electronics failures, with an in-lock percentage of 100% and with
resonators rarely going normal. Average resonator field levels used
during data taking runs are shown in table 1.

As can be seen, the three resonators in cryostat C, the first to
be installed, have the best performance. Two of the resonators in
this cryostat have fields that are limited by the ability of the
resonator control electronics to maintain phase lock. The relatively
poor performance of the resonators in cryostat B is probably due to
poorer drift tube venting due to differences in the drift tube vent
plumbing.

In addition to improving our method of drift tube venting we
intend to improve our conditioning techniques by the use of a larger
RF amplifier (we are currently limited to 200 W), and by installing
variable RF drive probes. We also intend to install the new loop VCX

units recently developed at Argonne[5]. Some of our resonators are over three years old and have not been rinsed since their acceptance tests. We are currently setting up a facility enabling us to rinse our resonators with trichloro-ethylene, as recommended by Argonne.

REFERENCES

1. F.J. Lynch, R.N. Lewis, L.M. Bollinger, W. Henning and O.D. Despe, Nucl. Instr. and Meth. 159, 245 (1979).

2. A.D. Frawley and J.D. Fox, Nucl. Instr. and Meth. 204, 37 (1982).

3. J.D. Fox and L. Wright, Rev. Sci. Instrum. 57 (5) 1986.

4. "Cryogenic Systems", by R.F. Barron, O.U.P., New York, 1985.

5. G. Zinkann, in these proceedings.

Table 1

Resonator		Average field level (MV/m)
B1	H45	1.33
B2	H46	1.75
B3	H47	1.23
B4	H48	Can-feedback cable shorted
C1	H36	2.04
C2	H35	VCX shorted
C3	H34	2.21
C4	H33	1.94
D1	H18	2.12
D2	H38	1.86
D3	H39	1.61
D4	H40	1.80

286

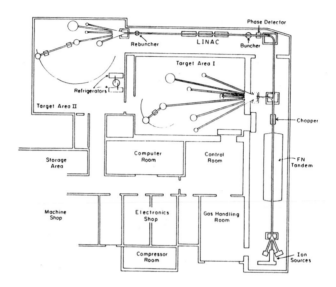

Fig.1   General layout of the F.S.U. heavy ion accelerator laboratory

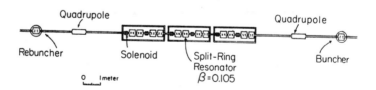

Fig.2   Schematic of the linac booster

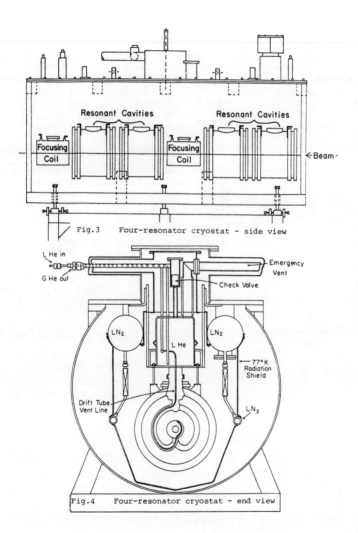

Fig.3   Four-resonator cryostat - side view

Fig.4   Four-resonator cryostat - end view

288

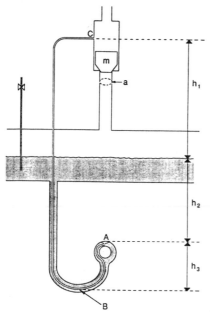

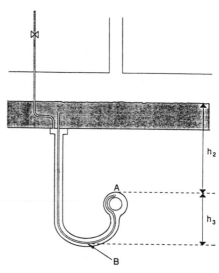

Fig.6   Schematic illustrating dewar filling
        through the resonator loading arms

Fig.5   Schematic illustrating choice
        of check-valve weight

Choose check valve so $\dfrac{mg}{a} = \Delta p_{c.v.} = (\rho_l - \rho_g) g h_1$

Discussion Following E. Myers Paper

STORM: Thank you. Do we have any questions?

PREBLE: When you were showing us those fields with the resonators, do your refrigerators support those?

MYERS: Oh, yes. When we turn the RF on the only difference it makes with regard to refrigeration is you just have to increase the fraction of time we run on precool.

So far we've only been running with one refrigerator, so it just increases the fraction of time you run with that refrigerator on precool. It's one of the good things or bad things about the Argonne resonators, is that in general you can't put much power into them before they go normal, so probably if the resonator is maybe four watts each, probably the additional load is only 30 watts, 30 or 40 watts. At the levels we're running at it's probably less than that.

GRAY: In a choice of your check valve you are aiming to pick a check valve so that the liquid stands all the way up to the check valve bottom, to the entry point?

MYERS: Yes.

GRAY: Is there a reason why you pick it so it stands all the way up there?

MYERS: Well, so that you can start the flow. The problem is if you were to make a check valve less heavy when you start off, the drip tube is presumably full of liquid, so in fact the whole line is going to fill with liquid. You force liquid down here through the line up here. So the thing is unless your check valve is capable of actually forcing liquid past that point you won't get any flow. The liquid head stops the flow of the gas.

The assumption is that this line remains full of liquid. We don't know that that's true, or we don't know that it's not true.

STORM: You have to assume the worst case because it will be true sometimes?

MYERS: Yes.

PREBLE: Do you know what temperature you were returning the liquid at? I mean, are you spitting liquid out warm, or is that a cold environment at the top of your relief tube, temperature wise?

MYERS: I'm not quite sure I heard your question. I'll try and answer it anyway.

Everything you see there is inside the nitrogen shields, we don't have separate thermometry, so we're not absolutely sure what the temperature at that point is. The assumption is it's still at four and a half degrees.

STORM: This would be the helium vapor temperature?

MAN IN AUDIENCE: It's a cold gas return?

MYERS: Right.

MAN IN AUDIENCE: You have that big FN transmitter. Do you still plan on using it for your high powered conditioning, or are you going to use something else?

MYERS: We are still planning eventually to use that large transmitter. Probably improvements which we should make with existing pulse conditioning system.

PARDO: Are those field levels that you showed really the realistic limits you can run for extended periods of time at this point?

MYERS: I should say we don't have the auto-start system operational right now. We haven't tried to get it operational. So those field levels, except perhaps one or two cases, are the field levels that can be sustained indefinitely, for a whole week, run without anybody having to do anything to the resonator.

Of course, they are significantly turned down from what you could achieve for an hour or so.

PARDO: That's not my question. My question is are those the best you can sustain for a week right now?

MYERS: For a week without anything going normal.

PARDO: And have those resonators been high power conditioned at all at this time?

MYERS: They have been high power conditioned. I think Gary saw our conditioning procedures, and probably it leaves something to be desired. So there is a question as to how effective the high power conditioning has been.

PARDO: Okay.

HYDER: What low-field Q values did you obtain on the resonators?

MYERS: Well, I guess I should say this is less relevant to the niobium resonators than to the lead-plated copper.

I think they are all pretty close to 10^9. It is a standard...

ZINKANN: A standard low-field Q, about one and a half to two, 10^9.

MYERS: Yes. The standard acceptance tests. They all met these standard acceptance tests.

GRAY: What is, typically your operating pressures, your dewar pressures compared to your return gas point pressure?

MYERS: Well, typically they run the dewar at 6 psi, there will be one and a half psi between the dewar and the pressure of the gas coming back from the Linac. So we have the one and a half psi drive pressure and gas back.

We have a return pressure on the refrigerator of about two psi, so that means it's two and a half psi drop across the refrigerator.

GRAY: Did your transfer system meet your delivery specs in terms of maximum liquid delivery rates?

FRAWLEY: It did not meet the heat loss spec. Every other spec. it was fine for. Pressure drops the flow rates. It does well.

THE CURRENT STATUS OF THE KANSAS STATE UNIVERSITY

SUPERCONDUCTING LINAC UPGRADE PROJECT

TOM J. GRAY
J. R. MACDONALD LABORATORY
DEPARTMENT OF PHYSICS, KANSAS STATE UNIVERSITY
MANHATTAN, KANSAS 66506

I. INTRODUCTION

The superconducting LINAC project is part of an upgrade in physics research capability at Kansas State University which also includes the fabrication of a cryogenic electron bombardment ion source, CRYEBIS, for the production of highly charged heavy ions. The upgrade is funded by a 5.1×10^6 grant from the U. S. Department of Energy, Division of Chemical Sciences. Additionally, the State of Kansas provided 1.09×10^6 for the additional new laboratory space required to house the new research facilities for ion-atom/molecule collisions studies. This upgrade project represents a significant advance in basic research instrumentation for physics studies involving interactions using fast heavy ions in areas other than nuclear physics.

The present laboratory space is $\sim$ 19,100 ft^2 with 9,100 ft^2 of that representing the new space built with state funding. Ground breaking for the laboratory addition was held in October, 1985, with building occupancy begun in January, 1987. The new space consists of accelerator/experimental space ($\sim$ 6000 ft^2) at basement level, access ways ($\sim$ 700 ft^2) at basement level, LHe refrigerator space ($\sim$ 1100 ft^2) at ground level, and shop/cryogenics repair space ($\sim$ 1300 ft^2).

Funding by DOE for the project began May 15, 1985. During 1985-86 design engineering on LINAC components proceeded and procurement of numerous subsystems was pursued. The laboratory organizational chart is given in Figure 1.

An average of $\sim$ 120 man hours/week in additional labor is provided by part-time undergraduates recruited from the physics and the engineering disciplines.

Listed in Table 1 are the major LINAC and Cryogenic Components, their current status, Vendor, and costs to date.

Table 1: Major LINAC Components

Component	Vendor	Status	Costs to Date
14 Nb Resonators	Argonne Nat'l Lab	Fabricated – under tests	$ 803,000
Cryostats	C.E. Raymond Manufacturing, a division of Combustion Engineering	Complete in-house	$ 120,000
LHe Refrigerator System LHe Distribution System	Cryogenic Consultants, Inc.	Installed, refrigerator and distribution system accepted	$ 646,000
Vacuum Pumps and related Vacuum Equipment	Leybold-Hereaus	In-house	$ 180,000

II. LINAC

The LINAC utilizes the superconducting Nb split-ring rf resonator cavities designed and developed at Argonne National Laboratory (ANL) for the ATLAS system. The KSU LINAC employs 9 low-β (β = 0.06) resonators and 5 high-β (β = 0.105) resonators. The resonators are fabricated using Stanford grade Nb at ANL and resonator performance characteristics, cavity Q vs. electric field gradient, are measured using the cryogenic resonator test facilities at ANL. Of the 9 low-β resonators, one is used as a time buncher (superbuncher) to prepare the beam from the tandem Van de Graaff for injection into the LINAC. The remaining 8 low-β resonators comprise the initial acceleration stages for the beam. One of the high-β resonators is utilized to energy rebunch the beam after acceleration by the LINAC while the remaining 4 high-β resonators comprise the final acceleration stage of the LINAC.

Tests on one low-β resonator, designated as L-15, were performed during May, 1986. Similarly, tests on one high-β resonator (H-52) were performed during July, 1987. The Q vs. electric field gradient for L-15 and H-52 are shown in Figures 2 and 3, respectively. Both Q curves exhibit Q values at low field levels in the mid 10^{+8} range. This is below the average Q values observed for similar resonators currently on ATLAS where $Q_{low\ E} \cong 10^9$. Resonator L-15 is scheduled for a washing procedure and retesting at KSU. Resonator H-52 has been disassembled and reelectropolished by ANL staff. Further testing during October, 1987 showed no marked improvement in the low Q measured in the low field limit. Additional testing of H-52 is scheduled for February, 1988. The design goal for the KSU LINAC is predicated on a maximum average operational electric field gradient of 3 MV/m. Both L-15 and H-52 would operate at this level without further improvement in resonator Q values. However, the type of degradation in resonator Q observed for L-15 and H-52 is indicative of surface contamination. Surface cleanup may improve overall resonator performance at a given input rf power level.

The remaining 12 resonators have been fabricated at ANL and final assembly and selective resonator testing was completed during October, 1987. Upon conclusion of these tests at ANL all resonators were shipped to KSU where further testing of the resonators will be undertaken.

The cryostats for the superbuncher (SB), energy rebuncher (ERB), and the three large acceleration modules (LCM1, LCM2, LCM3) have been fabricated per KSU design by C. E. Raymond Manufacturing, a division of Combustion Engineering, Enterprise, KS. The SB and ERB cryostats were electropolished and are inhouse waiting installation on their respective support stands. The large modules LCM1, LCM2, and LCM3 have been fabricated and were electropolished by Electroglo at their Elkhorn, WI facilities. The cryostats LCM1, LCM2, and LCM3 are in-house undergoing final vacuum tests. All cryostats were He-leak tested at the vendor's site by KSU staff prior to shipment. Only major weldments were checked at the vendor's site.

All UHV vacuum components for the LINAC have been acquired. The decision was made to use grease bearing turbomolecular high vacuum pumps because of the flexibility in mounting provided by such turbo pumps. Leybold-Hereaus model TMP360 turbo pumps were procured. Our experience with the turbo pumps to date is not acceptable. We have had a 62% failure rate on these pumps with 8 out of 13 pumps exhibiting bearing failure for operational time ranging from <200 hours to ~ 2300 hours. Our standard operational procedure is to put a pump into test operation upon receipt to avoid bearing flat-spotting conditions which may occur with inoperation, i.e., shell life. Manufacturing representatives claim that quality control in bearing mount machining procedures is the source of the bearing failures which we observe. We recommend awareness of the potential problems with high bearing failure rates in pumps of the type and manufacture with which we have had experience to date.

III. CRYOGENICS

The schematic of the LHe refrigeration system is given in Fig. 4. Table 2 lists the design specifications for the LHe/GHe distribution system as well as the measured performance levels as of October, 1987.

Table 2:

	LHe Refrigerator/Distribution		Design & Measured Maximum Performances	
	Refrigerator		Distribution System	
Design:	300w @ 4.2°K w/o LN$_2$	500w @ 4.2°K w LN$_2$	103ℓ/hr LHe per single cryostat	600ℓ/hr total delivery rate at cryostats
Measured:	337w	~660w	103ℓ/hr	607ℓ/hr

The LHe refrigerator and LHe/GHe distribution system was designed and fabricated by the prime contractor, Cryogenic Consultants, Inc. Our design specifications were predicated upon the necessity to meet several goals. These are:

1) Sufficient refrigeration capacity at the points of delivery to meet the LINAC cooling requirements with excess capacity in reserve.

2) Long term operation without the need for LN$_2$ precooling in the refrigeration cold box because of LN$_2$ costs involved.

3) Low loss LHe/GHe distribution system.

4) System reliability and field serviceability -
vendor support.

5) Overall system costs

The installation of the LHe refrigeration system was begun in
February, 1987. Shakedown and initial testing were begun in March, 1987,
with personnel from Cryogenic Consultants and KSU. Testing of the system
was interrupted randomly by problems encountered. The problems encountered were related to quality control situations involving the cryolab
cryogenic valving and valve tappet material hardness. Failure of the
exhaust valve tappet on one of the two expansion engines led to over-
heating of that engine and subsequent engine failure. Engine replacement
allowed testing to continue. During late August - early September, 1987,
all major system design specifications were tested with the results given
in Table 2. The refrigerator performance tests were conducted with and
without LN_2 precooling and the rated refrigeration levels were measured by
addition of a heat load to an insertion heater in the 1000ℓ LHe storage
dewar. This heater was modified with the addition of Cu cooling fins to
give a total surface contact area of 840 cm^2 with the LHe bath in the
dewar. At a LHe film boiling threshold of 0.8w/cm^2, we could readily
supply up to ~ 870 watts of heat load to the LHe bath. To date we have
tested refrigerator performance at the 1000ℓ dewar to 600w of refrigeration. In this mode the refrigerator was still taking warm gas from
storage and producing LHe in addition to providing 600w of continuous
refrigeration at 4.2°K. It is estimated that the real refrigeration limit
may be ~ 660w with LN_2 precooling. This is 130% of the design specification. Without LN_2 precooling a stable operational level of 337w continuous refrigeration @ 4.2°K was achieved. As our running experience to
date is somewhat limited, we may expect to possibly achieve a small
increase in refrigerator performance under these conditions once more
experience is obtained by the technical staff.

A system of test pots was installed on the ends of the LHe/GHe
distribution system. The test pots were fitting with 100w heaters, Si
diode temperature sensors, and 20 cm LHe level gauges. Measurement of the
total refrigerator system performance at the delivery point to the indi-

idual LINAC cryostats was made. Performance at these points is really
he figure of merit. Shown in Table 3 are sample test data accumulated to
ate. Also shown in Table 3 are the maximum expected heat loads for the
perational LINAC. The tests to date shown that the distribution system
loes provide LHe delivery capacity in excess of our needs for large
cryostats 1, 2, and 3. Without warm gas bypass on the buncher cryostats
he ends of the distribution system, we have met the projected LHe cooling
requirements. However, as of early September 1987, we could only get up
o 20 w per buncher cryostat under stable refrigerator running conditions
by using warm gas bypass from these two cryostats. Warm gas return
injects He into the compressor suction side of the refrigerator thus
lowering the overall refrigerator efficiency through loss of refrigeration
o the cold box provided normally by the cold gas return path.

TABLE 3:   LHe/GHe Distribution Test Results W/O LN$_2$ Precooling

CRYOSTAT		SB	LCM1	LCM2	LCM3	ERB	TOTAL
EXPECTED HEAT LOAD		8w	32w	32w	32w	8w	112w
SAMPLE TEST* RESULTS		0	103w	88w	74w	0	264w
		12w	70w	70w	70w	9w	231w
		20w**	38w	31w	39w	20w**	148w
	***	90w	88w	88w	88w	89w	443w

* tests made under stable refrigerator operational conditions
** warm gas bypass valves opened on these cryostats
***this test made using LN$_2$ precooling

Examination of the test pot mounting revealed that a virtual
thermal short between the inlet LHe line and warmer GHe return line was
responsible for the limited performance of the LHe/GHe distribution system
as far as the buncher cryostat positions were concerned. Extension of the
inner LHe delivery line was made so that the inner line penetrated into the
test pot volume in lieu of its original termination well up inside the GHe
return line. Flash loss of LHe at the entry point to the test pot in each
case was dramatically reduced thus giving a higher achievable cooling rate
for the two buncher positions on the distribution line. Design specifications called for 103 w of cooling maximum per cryostat at the ends of the

distribution system. The last entry in Table 3 gives the performance data for the LHe/GHe system. The system ran stable with 433w of input power to the 5 test pots, thus providing an equivalent of >600 ℓ/hr LHe flow in the distribution system.

IV. FUTURE

Overall there have been some gains and some slippages in schedule caused by rearrangement of priorities (gains) and vendor schedule delays (slippages). The final construction phase of the LINAC proper commenced during December, 1987. We anticipate meeting the schedule barring unforeseen delays. The previous work experience, and help of others at Florida State University, and ANL have been most valuable to us. We appreciate the exchange of ideas offered to us by our colleagues. We have recently made engineering design changes to handle the 2-phase LHe/GHe return flow problem encountered at Florida State University. Primarily, we have included an in-cryostat phase converter to insure that only GHe returns to the refrigerator coldbox. We have also redesigned the internal LHe plumbing to allow the option/combination of resonator spray cooling/gravity flow cooling.

This project is supported by the U. S. Department of Energy, Division of Chemical Sciences.

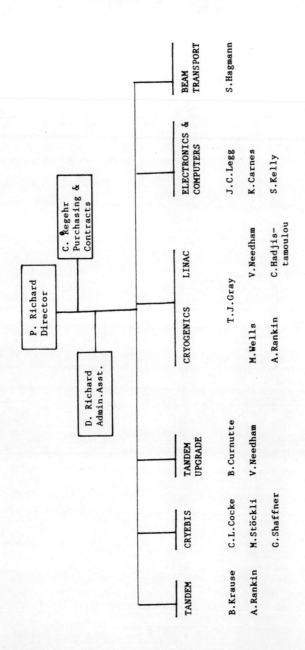

FIGURE 1: ORGANIZATIONAL CHART

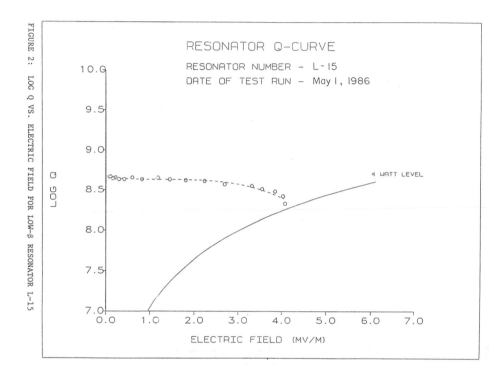

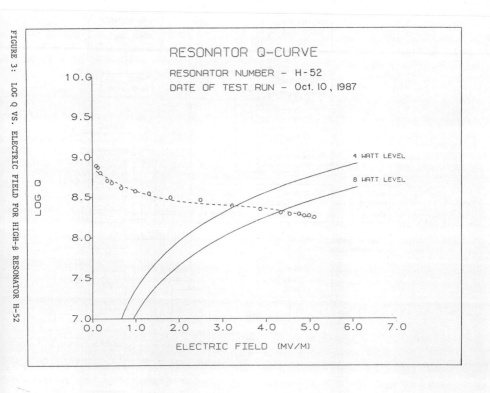

FIGURE 3: LOG Q VS. ELECTRIC FIELD FOR HIGH-β RESONATOR H-52

302

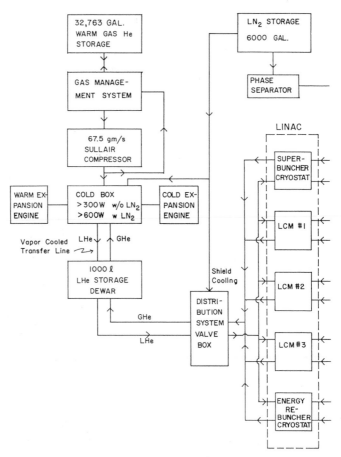

FIGURE 4: CRYOGENICS SCHEMATIC FOR THE LHe REFRIGERATOR,
LHe/GHe DISTRIBUTION SYSTEM, AND LN$_2$ SUPPORT SYSTEM

Discussion Following T. Gray Paper

STORM: Do you have any questions for Tom?

PARDO: The capacity numbers on your refrigerator I would assume are under pretty clean conditions. Have you gotten enough contamination in your system to observe how the traps plug and how that falls off, that sort of everyday problem?

GRAY: I personally haven't had that much experience with the refrigerator. I will simply tell you that starting up we took tube trailer gas, which was supposed to be of a certain quality and it was not. The neon content was in excess of 20 parts per million.

We got into a problem of plugging things right away with ice blockages. That gas had to be dumped and was replaced with higher purity helium.

I can't give you a total number of hours we have run the refrigerator, but we haven't gotten any serious problems with contamination since that time.

PARDO: Are your traps larger than the 2800 traps?

GRAY: Well, now when you say traps, the only thing we really have is a cool bed, which is on the refrigerator itself.

PARDO: That's what I mean.

GRAY: Okay, and I don't know exactly what the volume of that trap is. Mike might be able to answer that question.

WELLS: I would say it's probably four times greater at least.

STORM: Greater than what?

PARDO: Is this the 2800 size?

STORM: I didn't know there is a standard 2800 size. There are not many 2800s.

PARDO: You may be right.

WELLS: The only reason that I bring it up is just with the single refrigerator, it has a nice control capability. The one obvious drawback is the inherent recycling needed in contamination used in failure modes and so forth.

GRAY: Right. Well, the cold bed is isolatable from the refrigeration process and can be warmed independently so you can clean the cold bed.

That cold bed, it's in a canister that is seven feet long approximately and has two sections to it. Each section is about eight inches in diameter and about three feet long. That's each separate canister. I don't know what the volume is.

WILL: Where did you get the contaminated gas?

GRAY: We got contaminated gas out of Otis, Kansas. From Airco. They can give you good gas. We thought we were getting good gas. We didn't.

MAN IN AUDIENCE: Did they give you your money back?

GRAY: I don't know the answer to that.

WELLS: The gas actually came from Texas somewhere.

WILL: Did it come from the National Bureau of Mines?

WELLS: I'm not sure of that.

WILL: I guess just a word of warning. The National Bureau of Mines, their standard analysis procedure as reported by Charles Seitz, S-e-i-t-z, is in order in gas chromatography, to get peaks that are not folded over in, the peaks where the response is, they analyze 23 parts per million neon in the gas. That achieves the goal of simple analysis that's in their test gas, but it also means they cannot adequately analyze for neon below that level unless they go to special analysis for you. If you do buy from the National Bureau of Mines, be forewarned that you need to be picky about the neon analysis.

GRAY: We had planned originally not to buy tube trailer gas; but on the recommendation of the vendor we did. We have two 500-liter liquid helium transportable dewars, and henceforward we will take only liquid and put the liquid into the 1000 liter dewars and put it into storage as gas. We don't plan to have any more tube trailer gas around.

Our second load was okay. We have had no problems to date with contamination.

STORM: Any other questions? I have a question about the Q measurements. These Qs you are getting, they are substandard, are they measured at Argonne or only at Kansas or what?

GRAY: They are measured in the test cryostat facilities at Argonne.

STORM: These are made at Argonne. Are the usual Argonne procedures followed? Is Shepherd involved in these, or are you guys in this on your own?

GRAY: Well, the Argonne staff is involved. They look over our shoulders.

Vince Needham and I have been the people primarily involved in testing Q resonators, and we have tested Q resonators for Argonne and Florida State and for ourselves, and we have done, I don't know, eight, 10, 12 resonators, somewhere along in there, and these are the first two where we have run into these kinds of problems.

We did see definite signs of contamination on H-51 when we took it apart. Now there was a question about where it came from. We could have contributed to that ourselves. It's called evaporative Saran Wrap.

You do get electron beams out of those resonators, we have definite proof of that, burn holes in Saran Wrap, which wasn't supposed to be there, incidentally.

STORM: We have all burned some Saran Wrap.

SESSION XII

NEW FAST TUNING SYSTEM FOR THE ATLAS ACCELERATOR

Gary Zinkann

Argonne National Laboratory, Argonne, IL 60439

ABSTRACT

This paper will discuss field level limitations for on-line operation of the Argonne superconducting split-ring resonators. It also describes development of a new fast tuning system for the split-ring superconducting resonators.

INTRODUCTION

In off-line tests performed without tuning systems mounted on the resonator, the H-type (B = 0.16) resonator typically operates at 4.2 Mv/m with 4 w of rf power dissipated into the He. It has been observed that with the tuning systems installed, and operating on-line that the average maximum field level degrades to about 2.8 Mv/m with 4 w of rf power input. Extensive tests have been performed, both on line and off line, to try to identify the cause of this degradation. Several malfunctions have been uncovered and will be discussed in this report.

COUPLING PORTS

A small percentage of the on-line resonators were found to have hairline fractures on stressed areas of a cold-drawn niobium coupling port. After thermally cycling the resonant cavities many times from room temperature to 4.5 K over a period of several years, these stress fractures developed. The crack caused rf losses, lowering the Q of the resonant cavity and sometimes causing thermal instability.

Design changes have been made to eliminate the stress fracture problem. As the old-version resonators develop cracks, they are repaired and upgraded.

FAST TUNING SYSTEM

Figure 1a shows the fast tuning system originally developed for ATLAS.[1] Several key features should be noted. The fast tuner is operated at 77-90 K, it is cooled by liquid nitrogen, and is thermally isolated from the 4-5 K resonators. This is necessary because the rf losses into the fast tuner is about 100 W per resonant cavity. A portion of this loss is caused by rf eddy-current heating of the capacitively coupled probe. The probe is several square cm of copper at 80-90 K in an rf magnetic field of typically 45 Gauss. A second source of rf loss of a similar magnitude is the rf loss in the PIN diodes, the active element in the fast tuning system.

At the present, the field level in many of the on-line resonant cavities is being limited by the original fast tuning system. (See Fig. 1a.) The capacitively coupled probe tip is cooled by conduction through a beryllium-oxide insulator: on repeated thermal re-cycling there have been failures of the braze joint between the copper and the beryllium-oxide insulator. This failure causes the probe tip to thermally float until it cools by thermal radiation into the 4.5 K interior of the resonator. Also the PIN diodes used can only switch up to 5 KVA of reactive power. There are currently available PIN diodes with larger power capabilities.

Figure 1b shows a new fast tuning system currently under development. RF coupling is via an inductive loop, which can be cooled by conduction directly through the copper. The electronic elements are cooled by immersion in liquid nitrogen. Also a new PIN diode (Microwave Associates type MP4000) has been employed.

Early tests of the new system have achieved phase stabilization of an H-type (B = 0.105) split-ring superconducting resonator at accelerating fields above 6 MV/m, and switching reactive rf power loads of more than 20 KVA. This is four times greater than handled by the old design.

Rf losses are smaller in the new design, by a factor of two. Although the development work is not finished on the new fast tuning system, it's present performance exceeds that of the old system.

ACKNOWLEDGEMENTS

This work was done under the supervision of Kenneth Shepard with the collaboration of Gene Clifft and John Bogaty. I would also like to thank Tom Gray and Vince Needham, from Kansas State, for their help in testing the new fast tuning system prototype.

Work supported by the U. S. Department of Energy, Nuclear Physics Division, under Contract W-31-109-ENG-38.

REFERENCES

[1]Proposal for ATLAS (Jan. 1978) and Addendum (Dec. 1978). Copies of these detailed discussions can be obtained from the Physics Division, Argonne National Laboratory.

312

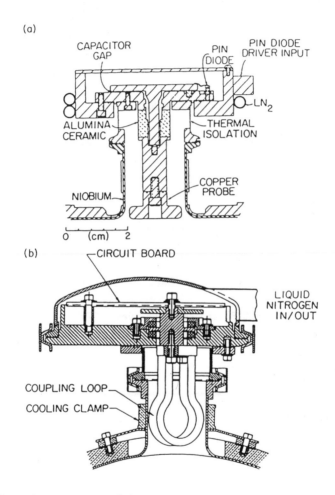

Figure 1. Two versions of the fast tuning system required to control the phase of ATLAS resonators. The first version (a) was capacitively coupled. Currently being deveoped is an inductively coupled version (b).

Discussion Following G. Zinkann Paper

GRAY: Thank you, Gary. Questions?

LARSON: You mentioned that with the old tuner you had about a quarter degree phase stability. Do you have a number for the new capability?

ZINKANN: I'm not sure how it scales with the switching rate. It could be proportional.

LARSON: Another question. You had this one anomalous test where you were up to at least 10 megavolts per meter. Has that been duplicated in any other resonator, or do you know why that happened to you?

ZINKANN: Yes, this resonator was the only one of its class, and I think the peak electric surface fields are comparable to what we have in our other resonators.

We were able to duplicate that result in that resonator on the order of three times over a period of a year, having it just sitting in storage.

LARSON: But I mean you haven't constructed another...

ZINKANN: And we won't construct another. That's the only one we need. But we, just to make sure it wasn't something funny going on, we let it sit, we ran it again, and it worked well.

STORM: I would like to know what you mean when you talk about a phase stability, because as I calculated you are running the phase back and forth by peak-to-peak amplitude of about two degrees, so what is this quarter degree? Is that the average phase, or is that the RMS or is that walk back and forth?

ZINKANN: I think it's the average.

STORM: I mean, the beam doesn't care what the average is. The beam goes through the machine during those excursions.

So it seems to me that you need to remember this tuning system is marching the phase back and forth, and then if you want talk RMS, say this is less, obviously with a triangular wave going up and down.

That may be a quarter. Certainly the peak amplitude is like one degree, so not so small.

ZINKANN: Yes. But that's why we increased the switching time of the diodes to help reduce that.

STORM: I can appreciate that, but it is important to make clear what the number mean.

MAN IN AUDIENCE: Could you tell me how high the RRR material is?

ZINKANN: The RRR material that we used based on some numbers that Ken Shepherd got from Hassan Pottamsay's paper he calculated to be between 150 and 200.

FRAWLEY: Can you tell me what the cooling was at six megavolts per meter for the tests you showed, best gas test?

ZINKANN: What the power was? On this one here the total power at 6.4 megavolts per meter, going into both the resonator and the amplifier, our fast tuner, was 84 watts. The power into the resonator was about 20 watts. So the power was quite high.

But you drop back a little from there, and it goes down substantially.

So, yes, it's more than we could do on line, because right now we're limited to eight watts of cooling, but that would still put us out in this area here, around five megavolts per meter.

CARNES: With the new VCX design you are relying on the gap capacitor or do you use an actual capacitor?

ZINKANN: Yes, we found some commercial capacitors that withstood thermal shock and the high RF powers that we were putting on them quite well, so our entire tuning system is going to be commercial capacitors that we use, and I didn't draw it, but the whole concept is different. It's two tuned circuits now, rather than lump capacitance, but I wasn't going to go into all that.

CARNES: When do you expect to have some of the drivers ready for production?

ZINKANN: Yes. I forgot to mention that. We will have six, we will have six modified VMOS FET drivers, if today's the first, they're supposed to be in our lab today. One of the things we did was we put a jumper in there so that we could run our present diodes with them, so we'll put them on line to switch our present diodes, and then when we modify we'll switch them over to switch the new diodes.

And we hope that just the new electronics in itself will eliminate a lot of our PIN diode problems. We've had a lot of failures of diodes, not from the RF problems or at the cryogenic end, but from the switching of the electronics.

GRAY: It's my perception that with the use of the RRR material, it's only used for the drift tubes and the loading arms? But the housing is still the Stanford grade material?

ZINKANN: Right.

GRAY: My perception though, it wasn't clear that the use of that material was going to give improved performance, but it obviously has, over what had been the experience with the previous resonators for the Stanford grade material used in the inner structure.

ZINKANN: Yes.

GRAY: Now there is a price differential the RRR material being more expensive. It's like 40 or 50 dollars a pound or something like that.

But our experience and I'm sure your experience is the major cost of construction and the fabrication of the resonator is not in the material, it's in the labor, and so if one is going to consider in the future constructing resonators I think that the decision to use an RRR material is certainly a valid one. One ought to look at that carefully.

I think it's a place where one might get substantial increases in potential field gradients with improvement of control circuitry and what have you as you have demonstrated.

ZINKANN: And in partial answer to Dan Larson's question, all of our new inner digitals were made out of the RRR material.

FRAWLEY: Are you planning to take one of your own high Beta resonators and do the same thing with it? I would be interested to see what one of the old ones without the RRR would perform like.

ZINKANN: When we get these four prototypes at the end of this month we plan to outfit them in a cryostat and see what they do.

I also neglected to mention that we took this resonator after the test run, that I showed the results with, and put it in our rebuncher slot and have been running it there. We haven't had much time on it, but it did run between three and a half to 4.1, and we ran that conservatively, so we didn't cause any problems and have a resonator to use for the experimental program.

FOX: Back to your zero. How do you cool the bond of the inner digital thing?

ZINKANN: This whole tube is just hollow.

FOX: How do you get helium down to it?

ZINKANN: It will hang in there, over the bottom. It is cooled by conduction through the end plate, through the niobium gasket. It's thermally tied to the end plate.

FOX: But it's just sheet niobium?

ZINKANN: Yes. It's solid niobium. It cools fine. There was no problem.

That was a consideration, was the cooling going to be sufficient, and it turned out it was.

GRAY: Are there other questions or comments?

STORM: The RRR material it looks to me like the low field Q was the same, and it didn't go into the field emission loss.

ZINKANN: That's right.

STORM: Is that the normal thing? I thought the RRR story was it didn't have the resistant loss. It cooled the resistant loss better.

ZINKANN: Right. We had hoped that what it would do for us was make any problem areas less severe, because it had a higher thermal conductivity.

I didn't know what it was going to do with the electron loading, and truthfully we didn't expect it was going to do much.

STORM: But is it correct to say that that seems to be the major effect?

My same experience experience and reading on the Q curve is you are not having that. The curve bends down like other curves but maybe not as much.

ZINKANN: No, it will electron load, but at a much higher field.

STORM: I think you're right. One thing, one problem you are finding is maybe it removes additional inclusions, and maybe you have better surfaces. That may be what we are sensitive to rather than cooling, because in fact the cool is hollow, so there is no real thermal transport except through the thickness of the material.

GRAY: I'd like to point out that with these kinds of field levels and this design of resonator, one is certainly approaching theoretical limits of the surface fields.

These surface fields are quite high on the magnetic fields we talk about, talked about a week or so ago at Argonne, fields in the range of up to 40 to 50 megavolts per meter surface fields. They are pushing it on up there.

ZINKANN: One more comment I'd like to add to Tom's comment about burning Saran Wrap, I did a test where I put a quartz window in the beam line of our accelerator and started conditioning resonators, and let me tell you, electrons come out. We charged up the quartz window until it arced over the ground. I could measure over six Rem of radiation, and when it arced over it, it just pegged the meter at over 100 Rem. A very narrow beam of high energy electrons.

So if you ever think of looking into the window of a resonator, don't do it.

GRAY: Before I relinquish the floor to Ken for announcements, this morning we have had the niobium resonators for you, it appears, with reports from Florida State and Argonne and Kansas State. This afternoon after lunch will be given over to the lead resonator camp. I think that certainly those areas are most interesting.

That's the other side of the superconducting Linac stories, and unequivocally they make nice progress in that area.

SESSION XIII

STATUS REPORT ON THE UNIVERSITY OF WASHINGTON SUPERCONDUCTING BOOSTER ACCELERATOR PROJECT

Derek W. Storm

Nuclear Physics Laboratory
University of Washington
Seattle, WA 98195 U.S.A.

I. INTRODUCTION

The University of Washington Superconducting Booster was proposed in 1982 and funded in December 1983. According to the original project schedule, acceptance was to take place in July 1987. The actual date of that milestone was late August 1987. The goals were to be able to accelerate particles from protons to middle weight ions using our 9-MV FN tandem as an injector. The design of this project and previous status reports have been presented at prior SNEAP meetings[1] and elsewhere.[2] Because we planned to accelerate lighter ions than other laboratories, we conceptualized the accelerator around "low-beta" resonators of β near 0.1 and high-beta resonators of some value from 0.16 to 0.2. The final plans were based on lead-plated copper quarter-wave resonators. We adapted a design of Ben-Zvi and Brennan[3] for our β = 0.1 resonator and designed a new β = 0.2 quarter-wave resonator.[4,5] In addition to the LINAC, we have built a 300-KV injector deck, which has our old direct extraction source and a new General Ionex 860 sputter source, and we have built a gridded buncher which runs at 50 MHz. These are fully operational.

Below we give a brief description of the LINAC. Then we discuss resonator construction and the rf control system. In the final sections we describe our operating experience and recent tests. Our most recent experience involved providing a ^{7}Li beam for the first nuclear physics experiment done with the new accelerator.

II.   DESCRIPTION OF THE LINAC

The booster accelerator consists of 24 β = 0.10 resonators in six cryostats and 12 β = 0.21 resonators in another six similar cryostats. Between each cryostat there is a quadrupole doublet. There is a β = 0.10 buncher and a β = 0.21 debuncher/rebuncher each in its own cryostat. Each pair of cryostats is pumped with a turbo pump. The beam from the tandem is shifted to the south with an isochronous transport system using four 45 degree bending magnets and two quadrupole doublets (the "dogleg"). Most of the beam line is pumped with turbo pumps, although two small ion pumps are used in the "dogleg" and a cryopump is used near the old HVEC switching magnet, where we expect a large water vapor load from the iron pole tips. The cryogenics system is based on a refrigerator built by Koch Processes, with a computer controlled distribution system which can supply liquid

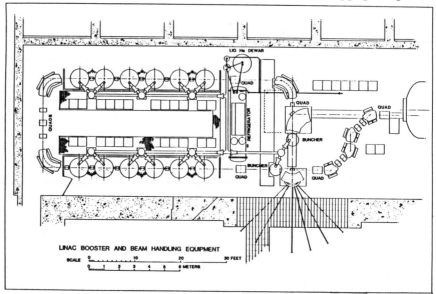

Fig. 1.   A plan view of the booster accelerator. The beam is bent through 180 degrees between the low-beta and high-beta halves of the accelerator, in order to utilize the space that was available in the Tandem vault.

helium and liquid nitrogen continuously to each cryostat. This system has been discussed at previous SNEAP meetings and will be discussed at this one.[6] A plan view of the accelerator is shown in Fig. 1.

The resonators and quadrupoles associated with a pair of cryostats are controlled by a single DEC Falcon computer. There is a local control box associated with each of these stations. Using buttons and a alpha-numeric video display, all the items can be controlled locally. The cryogenics and vacuum systems are each controlled by a PDP-11/23+. Each of these stations has a color display with a touch screen for local control. The injector deck

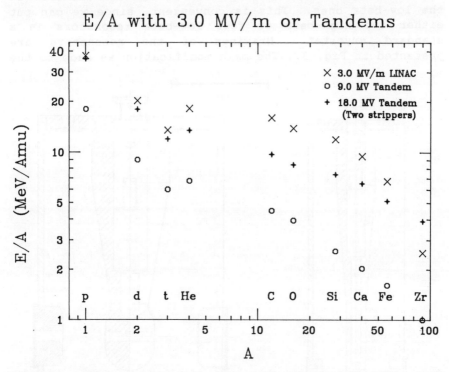

Fig. 2. Energy per unit mass for ions from the 9-MV tandem alone and from our LINAC injected by the tandem, assuming 3-MV/m fields in the LINAC and assuming most probable charge states following stripping in the tandem and between the tandem and LINAC. Energies for an 18-MV tandem with two strippers are also shown.

is controlled by a Falcon. All these satellite computers are connected to a DEC MicroVax II. This sytem has been described previously.[7]

The resonators operate at 150 MHz (actually 148.803). With an average accelerating field of 3.0 MV/m in each resonator, we would double the energy of the tandem proton beam and increase energies by about a factor of four for tandem beams of ions up to about mass 50. A graph describing the ion energies is shown in Fig. 2.

III. CONSTRUCTION OF RESONATORS

The high-beta resonators are double the diameter of the low-beta ones. This is convenient, since we can put either two high-beta or four low-beta resonators in a standard cryostat. Drawings of the resonators are presented in Fig. 3. The main modification we made to the

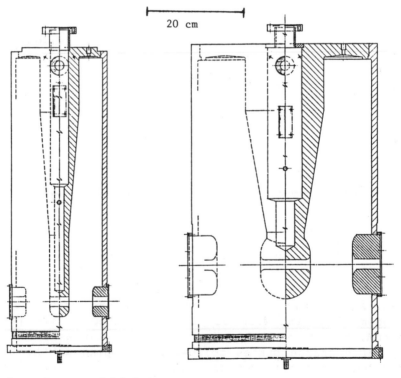

20 cm

Fig. 3. Low- and high-beta resonators.

construction scheme developed by Ben-Zvi and Brennan was
to make the inner conductor and shorting plate out of a
single OHFC forging. This simplified construction by
removing an e-beam weld step.

We built 29 low-beta and 13 high-beta resonators
between January 1985 and April 1987. The resonators were
lead plated and polished using a modified version[5] of the
technique developed at Cal Tech.[8] We did not try the
"modern"[9] thin plating technique during our production
runs, since the more tedious plate and polish technique
had been proven successful. Also, we had enough
experience with that technique to know that if the plating
looked good the resonator would perform well. The
resonator construction has been discussed recently.[10] The
resonators were installed in cryostats and brought on line
along with their control stations during the last two
years.

IV. RF CONTROL SYSTEM

Each resonator is operated in a self excited loop
with strong coupling. The quiescent power for the loop is
set to about 50 W for the low-beta and 100 W for the high-
beta resonators. Most of this power is reflected back
from the resonator. We use a circulator to direct the
reflected power to a 50-Ohm load. The amplitude of the
field in the resonator is adjusted to roughly the desired
value using the variable (inductive) coupler. An input
attenuator in the loop is adjusted so that its output is a
standard amplitude for the desired resonator field.

The frequency is adjusted using a differential screw
and lever mechanism to deform the bottom plate. This
mechanism has a range of about 10 kHz for the low-beta and
30 kHz for the high-beta resonators, and this range
requires 10^4 steps of a motor mounted on the top of the
cryostat. Tuning to within several Hz of the clock
frequency is possible. Once the resonators are tuned they
usually need to be retuned only if some large change in
amplitude or coupling is made.

Once the resonator is running at about the right
frequency and amplitude it is phase and amplitude locked.
The control unit was developed at the Weizmann Institute

and Stony Brook,[11] following a scheme of Delayen.[12] The output is developed using a "complex phasor modulator" built by Olektron Co. This unit permits independent control of the real and the quadrature power in order to control amplitude and phase independently.

V. OPERATING EXPERIENCE

Our resonator operating experience to date consists of a set of test measurements and conditioning runs with each resonator as well as four several day runs with beam and a fourth several day run with all resonators running locked. The resonators do not show signs of ageing; that is we can reproduce Q vs field curves, and once multipactor conditioning has been successfully concluded, we rarely see signs of multipactoring again, provided the resonators are stored at liquid nitrogen temperature under reasonably good vacuum. Some resonators have been in their cryostats for 20 months.

It is essential to perform helium conditioning or power conditioning in order to reach reasonable operating fields with our resonators. In Fig. 4 we present curves of Q vs field for the best high and low-beta resonator as well as for the median resonators. (The resonators were ranked by the fields we obtained during operation to determine the median.)

The resonators are extremely stable in eigenfrequency. Normal laboratory sources of mechanical vibration do not influence the phase error signal. The vibrations induced by banging on a cryostat with a board do make a noticeable disturbance on the phase error signal, but such activity does not cause the resonators to go out of lock. During tests of the prototype high-beta resonators we detected mechanical vibrations of the tuning plate, which is essentially a diaphragm with a fundamental frequency of about 500 Hz. Once the plate was supported in the center by the tuner mechanism, the vibrations were not observed.

We have observed a short term drift in the amplitude control when the resonator is first turned on. This drift appears to be due to the temperature coefficient of the detector diode and to heating of the control unit when the rf is turned on. When the resonator is turned off, we

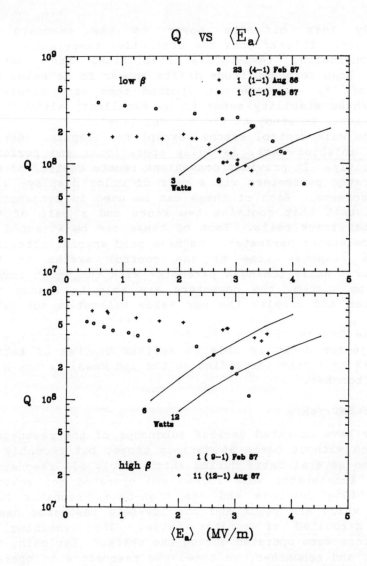

Fig. 4. Q vs Average accelerating field for the best and the median low- and high-beta resonators in the accelerator.

actually turn off the power to the on-board rf amplifiers. This effects the controller temperature. We are considering temperature stabilizing the control units. Long term amplitude drifts appear to be below the level of 1%. We have not studied them very carefully yet. Phase stability seems to be excellent, although it is difficult to study at the 0.1 deg level.

The main control system, except for computer control of the injector deck, is fully operational and performs excellently. It provides convenient remote control of all accelerator parameters via a pair of color displays with touch screens. Each of these can be used in conjunction with a unit that contains two knobs and a pair of two parameter track balls. Each of these can be attached to any accelerator parameter. We have paid special attention to the response time of the control system to the operator. While tuning a parameter with the touch screen and knob system, the computers are able to change the parameter and display the new value without a noticable delay.

The buncher before the tandem, in conjunction with the injector deck, is able to deliver bunches of better than 1/2 to 1 nsec (depending on the ion mass) to the high energy buncher.

VI. RECENT TESTS

We have operated various subgroups of the resonators with and without beams at various times, but recently we have had several tests during which nearly all resonators in the accelerator were locked and operated for several days. (One low-beta and one high-beta resonator have tuners which are stuck and one low-beta resonator has a short circuited rf monitor cable. The remaining 35 resonators were operated during the tests.) Excluding the buncher and rebuncher, we tuned the resonators to operate at about 6 to 8 Watts each for the low-beta and 14 to 16 Watts for the high-beta ones. Under these conditions we find the average low-beta resonator produces an average accelerating field of 2.83 MV/m, while for the high-beta resonators the value is 2.53 MV/m.

The distribution of fields is fairly symmetric, with a standard deviation of 0.34 MV/m for each type of resonator. The field values are determined for each resonator by measuring the energy increase it gives the beam. Except for one high-beta resonator, these are the units that were originally installed in the accelerator as it was assembled.

Our most recent two operating experiences involved the entire accelerator, using the resonators as described above. First, we were able to put an oxygen beam into one of our scattering chambers. By observing elastic scattering from a gold target, we determined that the energy resolution was better than 1% and the time resolution was better than 1/3 nsec. The beam was stable. We obtained 30% transmission from the entrance of the linac to the target. The resonators stayed locked at the fields mentioned above during the tests. Those figures, although not yet our design goals, represent a major step toward achieving a working accelerator. We expect that by replating the poorer performing resonators we should be able to achieve our design fields of 3.0 MV/m. (Note that this involves improvement of the resonator performance by 0.5 or 1.4 standard deviation, respectively, for the low and high-beta units.)

More recently, we have supplied a ^{7}Li beam for the first nuclear physics experiment with the booster. The energy was 88 MeV and the energy resolution was better than 1/2%. We are looking forward to continuing nuclear physics experiments as well as to fine tuning the accelerator.

References

1. Amsbaugh, J.F., et al, Rev. Sci. Instrum. 57, 761 (1986).
2. Storm, D.W., et al, IEEE Trans. Nucl. Sci. NS-32, 3262 (1985).
3. Ben-Zvi, I. and Brennan, J.M., Nucl. Instrum. Methods 212, 73 (1983); Brennan, J.M., Kurup, B., Ben-Zvi, I., and Sokolowski, J.S., Nucl. Instrum. Methods A242, 23 (1985).
4. Storm, D.W., et al, Proceedings of the Second Workshop on rf Superconductivity, Geneva, (1984) p. 173.

5. Storm, D.W., Brennan, J.M., and Ben-Zvi, I.,
    IEEE Trans. Nucl. Sci. NS-32, 3607 (1985).
6. Will, D.I., Proceedings of SNEAP, Stony Brook (1984)
    p. 273; Will, D.I., these proceedings.
7. Swanson, H.E., et al, Rev. Sci. Instrum. 57, 784
    (1984).
8. Burt, W.W., Adv. Cryog. Eng. Mater. 29, 159 (1983);
    Dick, G.J., Delayen, J.R., and Yen, H.C., IEEE
    Trans. Nucl. Sci. NS-24, 1130 (1977).
9. Delayen, J.R., Rev. Sci. Instrum. 57, 766 (1986).
10. Storm, D.W., Howe, M.A., Lin, Q.-X., and
    Rosenzweig, D.R., Proceedings of the Third Workshop
    on rf Superconductivity, Argonne, IL (1987), to be
    published.
11. Ben-Zvi, I., et al, Nucl. Instrum. Methods A245 1,
    (1986).
12. Delayen, J.R., Dick, G.J., and Mercereau, J.E., IEEE
    Trans. on Nucl. Sci. NS-24, 1759 (1977).

Discussion Following D. Storm Paper

PARDO: Any questions?

PREBLE: Are you having a hard time conditioning, where is it going?

STORM: Well, I don't know what you mean by hard time. We have the design from this that we used for the helium conditioning. We conditioned high or low, B resonators, with about one kilowatt pulse power at about three percent duty factor for a total period of about half an hour. We find during that time that the field that you can reach with that power increases substantially at first and then slower, and by the end of the half hour you've gotten about as far as you're going to get.

If we can get the resonators up to pulse of about seven or so megavolts per meter, then they will usually run pretty well after that.

If we can't, then I think they've got things other than field emissions limiting their performance, and that's not going to get fixed.

I'd like to be able to couple in a little bit more power. This amplifier has a tendency to go off with a loud bang if you don't treat it fairly well.

ZINKANN: Who's the manufacturer of your RF amplifiers?

STORM: RF Power. Yes, our neighbor in Woodenville, RF Power Labs.

CARNES: I think it was RF Power Labs. Do you have circulators on those?

STORM: Yes, they have circulators. They are set to trip on a 30 per cent reflective power.

FRAWLEY: Do you run any percentage out of lock while you are in the experiment that you remember? I have no feel for that.

STORM: Well, once you lock these resonators, unless something goes wrong, they stay locked. We did go out of lock, and one of the reasons was that we had a lying priced-out helium level indicator which the computer was trying very hard to get to read lower than it was reading, so it almost ran out of helium, but not quite.

So the thing didn't really want to warm up, but it wasn't getting enough helium to run superconducting with any power. It took us a few hours to figure that one out.

We had for a while some of the amplifiers that were tripping on over-temperature, so we put a fan on and things like that, but other than that, once we locked them they stay locked.

PARDO: So you do not see any evidence of occasional breakdown due to unknowns?

STORM: Well, I see some of that usually in the beginning when I get going with them, and then it stops. So far as vibration induced, the first was the material within the system, our buncher was taken up to a point, and I tried to see if I could tell by looking at it if there was anything. I didn't see anything. I took a two by four and beat on it and made it better, but I didn't find the answer.

ZINKANN: Forget the processing for the lead, but do you use solvents as a rinse at all?

STORM: The thing that we use I consider it an old-fashioned lead plate, and then polish. We plate with too much lead and then polish it off to get a shiny surface. We rinse it, the final rinse is with acetone, and we go to some pains not to leave a bunch of acetone around that dries on the surface.

ZINKANN: You don't use alcohol at all?

STORM: No, we don't use alcohol at all. I say it's old-fashioned. We do this because this is the old technique which we adopted with our prototype which worked, and we know that if we get a resonator which looks decent that it's extremely likely that it will perform well, and I have been very loathe to get into sort of a mixture of development and construction at the same time. I really don't want to have three cryostats with resonators that were made by some wonderful new process that then we have to take apart again.

ZINKANN: Do you have a development resonator you have kept on the side?

STORM: We have some spare resonators, and when we get spare time I certainly would like to try a lot of these things.

As I say, it's an old-fashioned technique. There's been a lot of development. A lot of it has been not very scientific or not very high statistics or not anything, and so, I think if we're going to do it we need to do it ourselves a few times until we know that it works.

I'm sure that there are better ways than what we do. We just need to find them.

Superconducting Heavy Ion Booster proposed for the JAERI Tandem

Suehiro Takeuchi
Japan Atomic Energy Research Institute
Tokai, Naka, Ibaraki Japan 319-11

1. Introduction

JAERI has a plan of a superconducting heavy ion booster for the JAERI tandem accelerator. The booster is assumed to comprise as many as forty niobium superconducting quarter wave resonators of beta = 0.1.

The quarter wave resonators are suitable because of their wide range velocity acceptance[1]. Ions of carbon to bismuth from the JAERI tandem can be efficiently accelerated by the resonators of only one beta of 0.1. It saves time and money to restrict the number of resonator types to only one.

The excellent result of a niobium quarter wave resonator at Argonne National Laboratory[2] was a motive for our development of niobium quarter wave resonators. If its maximum accelerating field gradient of 4.7 MV/m is realized with our quarter wave resonators the maximum accelerating voltage will be 28 MV. The energy-mass performance shown in fig. 1 was calculated for a stable terminal voltage of 16 MV. As a reasonable estimation for the operating condition, a field gradient of 3.3 MV/m will be possible and an accelerating voltage of 20 MV will be securely attainable.

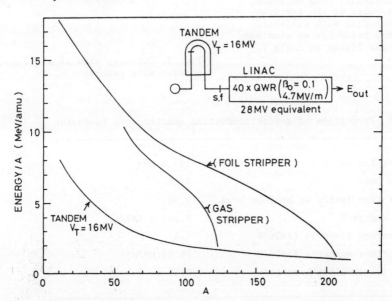

Fig. 1. Energy - mass performance of the JAERI tandem booster calculated with a terminal voltage of 16 MV for the cases of foil and gas strippers at the high voltage terminal.

We have been working on the superconducting quarter wave resonators since 1984. The results of the fabrication and the test of a prototype resonator are presented in section 2, the construction of superconducting buncher and de/rebuncher in section 3 and the booster plan in section 4.

2. Prototype Resonator

2.1 Design

The Argonne's quarter wave resonator presented a good result. However, further considerations were required, because the Argonne's resonator did not have beam ports nor frequency tuners. As a result of considerations on these points, we decided to have an oval cylinder for the outer conductor. The protype resonator is illustrated in fig. 2. Its properties, many of which were measured by the perturbation technique with dielectric and metal beads for an aluminum model, are listed in table 1.

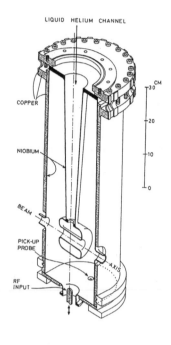

Fig. 2. Cut-away view of a quarter wave resonator

Table 1. Properties of a superconducting quarter wave resonator

Resonant frequency	129.6 MHz
Optimum beta	0.10
Transit time factor at optimum beta	0.90
Stored energy/E_a^2	0.046 J/(MV/m)^2
Peak surface electric field/E_a	4.6
Peak surface magnetic field/E_a	75 G/(MV/m)
Inside length along beam axis	0.15 m

* E_a = accelerating field gradient over the inside length.

2.2 Fabrication

The fabrication techniques of electron beam melting of niobium, explosive bonding and heat treatment were found to be available in domestic companies in 1984. After obtaining high purity(Stanford grade) niobium and explosively bonded niobium-copper composite materials in 1985, the prototype resonator was built by Mitsubishi Electric Company in Kobe(west of Osaka) in 1985-86. The center conductor components were made by cutting out thick niobium rods. They were welded together by electron beam welding in vacuum of 1×10^{-4} Torr, anneald at 1000°C for 6 hours in vacuum of 1×10^{-6} Torr, electro-polished by 120 - 150 μm at JAERI and annealed again.

The outer conductor was composed of U-shaped two halves made of niobium-copper composite plates. The weld lines of the two halves lay in the perpendicular plane of the beam axis. The inner niobium surface was electro-polished by about 120 μm. These two major components are shown in fig. 3.

At this step, we found weld defects near the upper end of the weld lines in the niobium outer cylinder. The defects were fixed, but the welds was found to be incomplete later. The center conductor part and the outer conductor were combined by electron beam welding to form the resonator. The resonator wall surface was electro-polished several times by 10-20 μm each to remove the small spots on the drift surface which were brought in by the second annealing process and to have smooth, clean and pure niobium surface as a final surface treatment. The resonator was finally rinsed with de-ionized water.

Fig. 3. Center and outer conductors of the prototype resonator before electro-polishing and final welding.

2.2 Test results

At the initial test, a maximum accelerating field gradient of 3.6 MV/m was obtained with a Q of 2×10^8. The result of Q values vs. E_a in fig. 4 was obtained after twice of surface treatments. The Q values are half as much as those of the Argonne's quarter wave resonator. The maximum field gradient of 4.6 MV/m with 1.6 watts rf input is high enough for us to have good future prospect.

Futher tests with the prototype resonator were tried. Outgassing at 75 °C for a weak was effective to improve the Q only at low field. Trials of further elecrto-polishing resulted in lower Q, a lower maximum field gradient of 3.3 MV/m and heavier electron multipactoring. After mechanical polishing of whole cavity surface and moderate electro-polishing, a 5 MV/m maximum field gradient was obtained, but the Q was not improved at all. The Q value at the maximum field was 1.5×10^8 and the rf power of 6.5 watts was required. The reason that the Q did not recover was probably due to the incomplete welds mentioned above. We expect Q values as much as 1×10^9 and a maximum accelerating field E_a of about 5 MV/m for a resonator without defects.

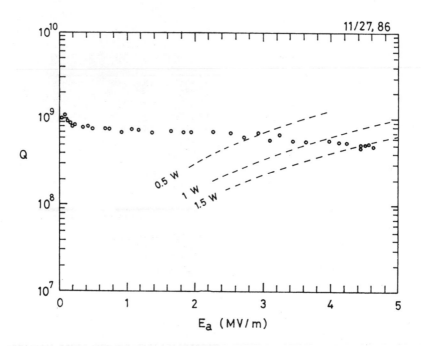

Fig. 4. Resonator Q vs. accelerating field E_a measured with a prototype quareter wave resonator at 4.2K. The broken lines show Q required to obtain a give field at a fixed power.

3. Superconducting post buncher and de/re buncher

Two superconducting quarter wave resonators and a cryostat have been under construction for the post buncher. Two sets of the center and outer conductors before the final electron beam welding are shown in fig. 5. The resonators were modified at several points. The outer cylinder was made more flexible for course and fine tunings. Wall thickness of the base end plate at the top was increased to pass the domestic high pressure gas regulations. It was required to measure the mechanical strengths of the niobium materials at the room and liquid helium temperatures. Welding was inspected carefully by a high pressure gas security agent. A leak test and a high pressure test were also required. After this symposium, the final surface treatment and performance tests will be done.

We will make a unit same as the buncher in a year for a prototype linac unit to test the resonators with beams. It will be used for the de/re buncher of the booster system in future.

Fig. 5. Center and outer conductors of two quarter wave resonators for a superconducting buncher before final electron beam welding.

338

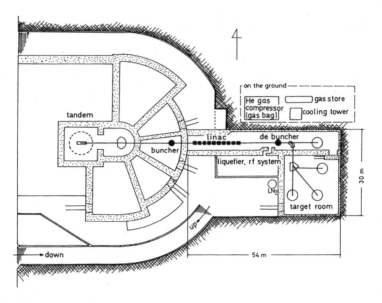

Fig. 6. Layout plan of the JAERI tandem superconducting heavy ion
booster. The left half of the figure is the existing building of the
tandem. The right hand side is planned.

4. Booster plan

A layout plan of the JAERI-tandem superconducting heavy ion booster
is illustrated in fig 6. We considered an idea to return the heavy ion
beams accelerated by the booster to the entrance of the beam switching
magnet of the tandem to use the existing target beam lines and
experimental equipments. We gave up the idea because of great
difficulties in the partial re-construction of the building of the
tandem. The plan of fig. 6 is simple and economical. It provides a new
target room to users of the tandem-booster. Above all, it is
advantageous that a straight beam transport does not degrade beam
properties.

Forty resonators will be housed in ten cryostats. Each identical
cryostat unit consists of four superconducting quarter wave resonators
of beta= 0.1 and a forcussing lens. The lens is either of a super
-conducting solenoid or a normal magnetic quadrupole doublet sitting out
side.

With respect to the beam bunching, pre-bunched beams of very heavy
ions heavier than a mass of about 60 may spread too much in a long drift
space of 8 m between the the terminal stripper and the high energy
acceleration tube entrance to be accepted by the post superconducting
buncher. It is not practical to re-bunch the beams in the high voltage
terminal. Then, in future, we will have to replace one of the two
bunching resonators by a resonator with the resonant frequency of 2 x
129.6 MHz in order to expand the phase acceptance approximately from 60°
to 180°.

From this point, the pre-tandem buncher is less useful for very heavy ions. We have other difficult situations against pre-bunching in the JAERI tandem. Probably, it will be hard to have good and stable pre-bunched beams from the tandem. Rather than struggling with the pre-bunching system, it would be better to use the tandem as a dc beam injector. It makes things quite simple. We are going to investigate quantitatively this problem, further.

A 350 watts helium requefying system, a rf control sytem and other components associated with the cryostats and so forth will be placed near the booster linac as are shown in fig. 6. We are proposing the construction of the booster beginning in 1988. The plan has been endorsed by the Science and Technology Agency of Japan. We are waiting for the approval of the Finance Ministry.

References:
1) Ben-Zvi and J.M. Brennan, Nucl. Instr. and Meth. 212(1983)73.
2) K.W. Shepard, S. Takeuchi and G.P Zinkann, IEEE Trans. Magn. MAG-21(1985)146.

Discussion Following S. Takeuchi Paper

PARDO: Any questions?

LARSON: I should think that a considerable part of your timing problem after the tandem is the 90 degree analyzing magnet rather than the drift space through the tandem.

TAKEUCHI: We are going to have DC beam here, so we have no problem. We are not going to install the prebuncher.

LARSON: Okay. I misunderstood.

TAKEUCHI: But we are worrying about the larger longitudinal emittance with not having a prebuncher.

PARDO: Yes, that, and I wonder about now your capture will not be as efficient. So you now have to worry about the unbunched beam in between.

TAKEUCHI: Right. So we are going to, as many resonators with different frequencies, with different harmonics to get wider acceptance for the bunching.

PARDO: But if you don't bunch at all before your high energy bunching your best efficiency will only be 70 per cent. 30 per cent of the beam will still be somewhere else.

TAKEUCHI: Right. We have to dispose of the rest of the beam.

FRAWLEY: We run with a 90 degree magnet, and we have no trouble with our longitudinal bunching for times like half a millisecond. That was my comment.

My question is what kind of a company did you have make the resonators?

TAKEUCHI: Mitsubishi Electric Company.

ZINKANN: Could you tell us the cost of those prototype resonators?

TAKEUCHI: We spent about $150,000.

ZINKANN: Would you expect that cost to come down significantly if you went into production?

TAKEUCHI: Not so significantly. We need more than $100,000 for a resonator.

PARDO: Was that only construction cost, or was that including electronics?

TAKEUCHI: I don't know.

STORM: I just had a remark. If it's hard to tune the thing by squeezing it, and it looks like it might be, it came to mind a way to tune it would be to make the bottom plate of thin copper like we do and either lead plate it, you have low current there, or see if you can sputter coat it with niobium. That might be a workable possibility.

TAKEUCHI: We didn't try the slow tuner yet, so I am worried about that, how it works.

STORM: It looks hard to squeeze it. The other way I suggested would work. It wouldn't be as neat, but I don't think you really need a pure niobium bottom. So those are other alternatives might work.

TAKEUCHI: Yes. We are considering the bottom part as our alternative part. We already tried slow tuning with the bottom part, which is made of niobium copper explosively bonded.

STORM: But it's thick.

TAKEUCHI: But I found that that is no good, because with respect to the difficulty in the fabrication. So we gave up that solution.

ZINKANN: Do you have an idea of what your frequency change is per .001 inch when you squeeze the sides?

TAKEUCHI: Frequency change is about 140 kilohertz per millimeter.

PARDO: 40/1000ths of an inch.

ZINKANN: Yes. So it's possible to do the slow tuning?

TAKEUCHI: I think it's possible by using the bellows and the helium gas.

LARSON: I'll make one further comment. I assume you are familiar with the double drift buncher concept, and you would need only two frequencies to achieve that with harmonic bunchers, instead of trying to use a larger number of cavities for your buncher system. Just two properly spaced is equivalent to at least three otherwise.

PARDO: So isn't that a double drift buncher that you are showing there, the two quarter wave?

TAKEUCHI: Yes. We are going to use the two quarter wave postbuncher and planning to exchange one of them with the higher frequency quarter wave resonator.

LARSON: But the secret of double drift bunchers is spacing them properly. There was no indication on the slide that that was done.

PARDO: I understand that's true, but I would assume that would have to be done. Just arbitrarily putting them there won't get you very far.

FRAWLEY: How much refrigeration capacity are you planning?

TAKEUCHI: We are considering 350 watts refrigerator.

SESSION XIV

UNIVERSITY OF WASHINGTON LINAC CRYOGENICS:
CONTROL, PERFORMANCE, MAINTENANCE, AND SAFETY

Douglas Irle Will, Research Engineer, Cryogenics Task Leader, University of Washington Nuclear Physics Laboratory, Mail Stop GL-10, Seattle, WA 98195.

David T. Schaafsma, formerly Research Engineer at UW, now a graduate student, Department of Physics, Brown University, Providence, Rhode Island.

John A. Wootress, Accelerator Technician, Cryogenics, University of Washington Nuclear Physics Laboratory, address above.

I. OVERVIEW

The University of Washington Superconducting Booster cryogenic system consists of the following items: an 8000 gallon MVE liquid nitrogen tank with 9 spools of Cryenco 1" inner line vacuum-jacketed delivery pipe; a Riley-Beaird 2664 ft^3 actual volume helium gas buffer with 250 psig working pressure; a Koch Process Systems 2830S with wet expansion engine and three RS screww compressors providing 500 watts nominal cooling at 4.3°K without liquid nitrogen assist; a Cryofab 1000 liter liquid helium dewar with 6" neck; a Beechcraft, Boulder Division, cryogenic distribution system between refrigerator and cryostats with active vacuum pumping and flanged service access providing valve controlled liquid delivery for nitrogen, plus liquid delivery and cold vapor return for helium; demountable field joints between delivery system and cryostats; fourteen Janis Research cryostats, including one buncher, twelve accelerator, and one rebuncher/debuncher units; a water cooling tower system with isolated deionized process water loop; and a Digital Equipment PDP 11/23+ with touch screen color monitor, analog and digital I/O, and remote control via the booster MicroVAX.

II.   CONTROL SYSTEM

The original goals for control of the cryogenic system have largely been achieved; later additions are close to implementation. The eventual objective is to create a cryogenic system that is largely self-monitoring and self-regulating.

Status

Currently, a dedicated cryogenics PDP 11/23+ microcomputer regulates liquid cryogen levels in 14 cryostats by changing the settings of 28 valves (one LHe and one LN_2 per cryostat) to maintain constant level sensor readings. Each control loop has variable setpoint and proportional closure range. The valves can be run in manual mode as well as in the automatic loop.

The remaining two original goals, the pressure building loop and the dewar heater loop are close to completion. The pressure building loop will throttle the dewar vapor return to maintain sufficient pressure to drive liquid and vapor helium to and from the cryostats. The dewar pressure sensor and the stepper motor to drive the pressure building valve are both installed but await cabling and testing. The dewar heater loop will drive the dewar heater to maintain constant dewar level sensor reading. The dewar level sensor, heater, and heater power meter are all installed. The power meter remains to be tested.

During initial operations of the above control system repeated computer lockups occurred. These were traced to electrical power system transients of short duration (one or two cycles). To eliminate lockups an uniterruptible power supply (UPS) now provides roughly eight minutes of computer backup power. In addition a circuit was installed to monitor the computer's heartbeat signal and fail critical actuators in a safe mode in case of lockup.

As a final backup an autodialer has been installed to dial the project director and the cryogenics engineer whenever a failure occurs. This autodialer is currently activated only by power outages. Eventually, other conditions will probably be monitored and used to trigger this device.

Equipment: The control system consists of a PDP 11/23+ microcomputer including various I/O boards, two monitors and a touch screen, numerous signal interface and isolation boards, the uninterruptible power supply, and various transducers. The PDP 11/23+ comes on a quad board which is mounted in the first of two MDB MLSI-BA11-2000-F crates. Its 1Mword RAM is half populated at present providing 512kwords on a dual board.

The human interface is via an Aydin color monitor with Carroll touch screen driven by a Peritek color board

which provides screen RAM residing in memory. Additional
active interaction is provided by similar touch screen
monitors via an RS-232 link to the Booster Control
microVAX in the control room. The RS-232 link allows
remote status display, remote adjustment of parameters
(setpoints, etc.), logging of cyogenic status onto disk,
and an independent malfunction alarm via the linac
control MicroVAX. A secondary link to the main VAX
permits print out and telephone modem access. Aydin
color monitors equipped with Carroll touch sense finger
position through x and y arrays of infrared LED's and
receivers. The work well if the arrays and their plastic
shield are cleaned quarterly. Two black and white
screens (driven by a similar Peritek B/W board) provide
status only information at remote locations.

Analog I/O comes in through eight ADV11C multiplexed
adc's and goes out through nine AAV11C dac's. Binary
status and control occupy two DRV11J parallel
interfaces. The multiplexed adc's allow choice of two
input modes: 16 channel single-ended, and 8 channel full
differential input. In either case the maximum input
signal is ± 10.5 V (signal plus common mode with respect
to computer ground). An attempt to use the differential
input mode to provide isolation from ground loops failed
when signals ≤ 9Vdc plus common mode noise exceeded 10.5
V magnitude. Table II-1 lists the items monitored.

Table II-1:   Items Monitored by Computer

Items Monitored for Control Purposes

LHe levels (14 cryostats and 1 dewar)
LN_2 levels (14 cryostats)
Dewar pressure (1)
He gas buffer pressure (1)
Differential pressure across distribution system vacuum
gate valve (1)
Distribution system vacuum Pirani gauges (2)

Items monitored for status information only

Dewar temperature, and heater current and voltage (3)
J-T valve and dewar pressure-building valve position
potentiometers (2)
He gas buffer temperature (1)
LHe distribution system differential pressure (1)
LN_2 tank level (1)
Distribution system vacuum ion gauges (10)
     (5 pairs of mantissas plus exponents)
Field joint Pirani gauges (14)
Cold box temperatures (7)
Expansion engine rpm's (3)
Refrigerator supply, J-T, and return pressures (3)

Signal conditioning is provided by the four lab-made
crates mentioned above. Buffer amplifiers drive cryogen
valves from dac's. Multiplexed logic decoder/amplifiers
drive stepper motors. Relays are used where necessary to
prevent ground loops and eliminate incoming spikes and
RFI. Analog output amplifiers drive power equipment such

as the dewar heater and stepper motors.

Future Plans

The PDP 11/23+ will monitor the ON/OFF and LOAD/UNLOAD status of the three RS helium compressors once cabling is completed. Each compressor will have an automatic startup relay installed to permit an operator to enable one or more compressors to restart automaticly after a power outage. Each will also include a time-delayed automatic load relay to ensure optimum running efficiency. (Unloaded screw compressors are less than half as efficient as loaded ones.) With three compressors capacity will be cut back by turning one or two off completely. The Koch refrigerator will also be monitored by the microcomputer. For each expansion engine status, speed, and inlet and outlet temperatures will be monitored.

III. PERFORMANCE

Liquid Nitrogen Consumption

Liquid nitrogen use at the nuclear physics lab is dominated by cryostat shielding. Experience with lead-plated copper resonators implicates moisture and organic compounds as degraders of resonator performance. We therefore decided early in the booster project to forgo all organic material (except the top flange seal O-ring) in our cryostat vacuums. With time this resolve has softened enough to allow small amounts of Torr Seal^{TM} used as an adhesive. We have avoided aluminized mylar insulation completely <u>in cryostat vacuums</u>. Consequently our fourteen cryostats consume from two to three liters per hour each (typically six square meters of shield surface each). Total cryostat consumption of roughly 250 gallons per day accounts for over 70% of the typical daily consumption of 350 gallons. Table III-1 summarizes our liquid Nitrogen use.

TABLE III-1:   Liquid Nitrogen Use

Item	Consumption
Quiescent	
All 14 cryostats	1000 liters/day
Distribution system	200 liters/day
Tank loss	100 liters/day
Exposed lines (improvements planned)	≈200 liters/day
Gas use, variable	≈200 liters/day
Subtotal	≈1700 liters/day
Active	
Power dissipated, 38 coupler intersepts	200 liters/day
Refrigerator precool (not normally used)	(≈700 liters/day)
Subtotal w/o refrigerator precool	200 liters/day
Totals w/o refrigerator precool	≈1900 liters/day

Liquid Helium Heat Budget

To those of you trained in traditional cryogenics where every watt of heat leak is tracked down and eliminated, our cryogenic system may seem wasteful. From an historical perspective, however, our particular path is quite logical. When we began design and purchase of the cryogenic system, our lab had no experience in liquid helium cryogenics. Discussions with engineers operating other systems led us to specify a refrigerator with twice the cooling capacity we estimated necessary (400 watts instead of the 200 estimated). We knew the refrigerator would need maintenance and might not be capable of peak capacity at some times. Therefore, our invitation to bid included a competitive advantage for any system capable of up to 600 watts. We also offered an advantage for ease of future expansion. The 2800 System bid by Kock Process Systems offered 600 watts peak cooling as is. It can be expended to 1000 watts capacity with minor engine changes plus two additional RS compressors. Thus, we knew in early 1984 that each additional 200 w of cooling cost less than $100,000 once we had this refrigerator.

Timely performance was critical and our schedule was tight. Discussions of distribution systems clarified that no well-refined, modular, efficient system existed commercially. Experienced cryogenic engineers felt that a twenty to thirty watt distribution system was possible only with careful refinement. A seventy watt system appeared possible with modest refinement from an initial design. Stony Brook's experience suggested a 150 watt system was likely if we failed to require an acceptance test <u>at the manufacturers plant</u>. We bid a fifty watt system with seventy watts allowable __AND__ required acceptance testing at the plant. Acceptance tests cost us an additional $100,000 on the bid __BUT__ these tests led to several improvements which reduced first test heat leaks of ≈300 watts to the final test heat leaks if ≈70 watts. (Again, roughly $100,000 for 200 watts). See Table III-2 for a summary.

TABLE III-2:   Heat Budget at Liquid Helium Temperature

Sources of Heat

Source	Heating:	Estimated	Actual
Quiescent			
Cryostats	all 14 combined	14 watts	30 watts
Distribution system		30 watts	70 watts
Active			
Power dissipated	all 38 resonators	153 watts	340 watts
Totals		197 watts	440 watts

Refrigeration

Conditions	Cooling:	Specified	Offered	Delivered
3 RS compressors w/o LN$_2$		250 watts	440 watts	480 watts
3 RS w/ LN$_2$ precool		400 watts	560 watts	570 watts
With one unit down for maintenance				
One RS down w/ LN$_2$ precool		≈300 watts		
One expansion engine down w/ LN$_2$		≈300 watts		

Balance w/ 3 RS w/o LN$_2$: excess cooling   40 watts

Now we have a reliable system running. The linac and its cryogenics were completed roughly on schedule. The system has redundancy with one expander or compressor down for service. The system is expandable. The linac is able to run _and_ there is now time to track and eliminate heat leaks.

IV. MAINTENANCE

Scheduled Maintenance

Most routine maintenance is directed at rotating machinery on our helium refrigerator and its compressors. The distribution system also requires minor service. Such maintenance is summarized in Table IV-1.

TABLE VI-1: Routine Maintenance, through September, 1987

Item	Hours to Date	Service	Frequency
Refrigerator Cold Box			
Top expander	15,000 @ ≈150 rpm	main seals	6x
		wrist pin bearings	2x
		valve seals	1x
		cam follower bearings	1x
Middle expd'r	10,000 @ ≈150 rpm	main seals	3x
Wet expander	2500 @ ≈70 rpm	main seals	2x
RS Compressors			
RS 1	12,000	replace charcoal	1x @ 5kHrs
RS 2	12,000	replace charcoal	1x @ 10kHrs
RS 3	12,000	replace charcoal	1x @ 10kHrs
Distribution System		He into vacuum	
	10,000	warm to derime	4x

Several comments are in order. The cold box was in use (with at least one expansion engine turning and at least one RS compressor on) 90% of 1986 and 99% of 1987. Note that we change compressor charcoal less frequently than specified by Koch. In the future we plan to replace it once a year. Our compressors are equipped with a new, efficient filter consisting of four Balston units in series before the charcoal bed. They have needed NO additional oil. The oil has a distinct smell and contamination appears quite evident if found. The worst contaminated bed showed several inches of visible oil but was clean to smell less than one foot from its top after 10,000 hours. This particular contamination resulted from our mistakenly overfilling the compressor once. The last bed changed had no evidence, visual or olfactory, of any contamination.

Incidental Maintenance

Cold box: We are inserting bronze wrist pin bearing sleeves into the aluminum crossheads (as a particular engine comes due for routine service) to prevent gouging of the aluminum when bearings are changed. One wrist pin needle bearing cup shed its cap into the sliding region between crosshead and guide, severely scoring both. Both were replaced at the time and we have since repaired the damaged items for spares.

R S compressors: None of the three RS compressors installed would pass the specified water flow given by Koch. These have not been repaired as Koch has informed us that the flow and temperatures at which we are currently operating are still acceptable.

Distribution system: We installed proper low pressure relief valves in all field joint bellows after a plug type pumpout/relief combination failed to relieve at low enough pressure.

Cryostats: Swash plate follower bearings (on the vacuum side) of several rotary feedthrough assemblies have broken their outer races. Subsequently all tuner (high torque) feedthroughs have since been fitted with sturdier needle bearings before installation. Annular liquid nitrogen tank design has been improved to prevent inner wall collapse, and strict precool procedures were instituted with special equipment to prevent collapse of those thin-walled tanks still in service. Two coaxial cables carrying radiofrequency power to resonators and located inside the cryostat vacuum space have required replacement. Two liquid helium level sensors have developed ground faults inside the cryostats. One has been repaired by isolating its controller; however, the other required a temporary sensor installed through the top auxilliary fill port.

V.   SAFETY

The general public's concern for the "radiation" hazards of a nuclear facility such as our lab is understandable (if somewhat misplaced) in this post Chernobyl era. This fear is ironic since many Americans worry little over the large annual lung-dose acquired by smoking a pack of cigarettes a day. What I find astounding is the anxiety evidenced by our physics graduate students when radiation protection is occasionally discussed in our weekly lab-wide meetings. Mention of the smallest discrepancy in our radiation protection system can easily consume half an hour with the ensuing discussion. In light of this deep concern, one can ask how serious this danger really is.

Historically, this lab has never had a single exposure over the weekly maximum allowable, 300 mR, in nearly 40 years of operation. Even our most senior employees have total accumulated exposures at or below 1% of their allowable totals. Since our cyclotron ceased operation two years ago, exposures have been even lower. In fact, no employee here has a an on-the-job accumulation greater than one tenth of his lifetime background accumulation. Again, ironicly, attempts to identify the real dangers of working at our facility pass almost unnoticed.

For the sake of comparison, let's group hazards into three categories: 1) those unlikely to cause serious injury or damage, 2) those capable of serious injury or damage, 3) those which are potentially lethal. Of course a proper risk assessmant would be more sophisticated, but this crude grouping will suffice to make my point.

Category One

At our lab penetrating radiation belongs in this category. An alternate method of assessing this danger is to ask what effort a significant single exposure or accumulation might require. Two forms of moderately intense radiation exist at the nuclear physics lab. Roughly one Rem per hour of neutrons can be found at some beamline locations during our hotest deuteron runs. Similarly, about one Rem per hour of ≈300keV X-rays can be found near our booster cryostats. Of course both areas are properly restricted at these times. A lethal accident would require violating an interlock AND draping oneself over a beam line or cryostat for most of the two thousand hours in a work year. In fact this would probably not suffice since radiation levels from most of our experimantal runs are so low that the general public could tour any and all parts of our tunnels and caves during operation. Conservative safety barrier placement, reinforced by monitors which are extremely loud and obnoxious, make it unlikely that barriers will be disregarded and essential impossible to exceed the maximum allowable weekly exposure in a weeklong hot run. The infrequency of hot runs and typical work patterns practically ensure that no one will accumulate a working exposure greater than the typical background exposure in Seattle. And as I mentioned before, typical accumulation by a worker in a year's time at NPL has been one tenth that background.

Category Two

The list of items capable of serious injury or damage includes the following: chemical exposure, falls, and machine tool injuries. While any of these has the potential for serious injury, actual incidents have been mostly minor. Anaerobic pipe joint compound has caused several skin rashes. Machine shop accidents have been limited to minor cuts and bruises with the exception of one fingernail which took a while growing back and one badly bruised foot (now recovered).

Category Three

The third category of potentially lethal accidents is the one I really want to address in comparison with the hazards of radiation. This group includes electric shock, the handling and use of compressed gases, and lifting of heavy objects with cranes (or other lifting devices.) All of these hazards have substantial potential for a lethal accident.

Elictrical shock: Adequate care in equipment design and construction have limited electric shock incidents to nonlethal circumstances. All potentially lethal power voltages are handled according to the National Electrical Code. High voltage direct current situations are considered carefully. If a power supply (or a static discharge) can produce a significant fraction of the lethal current (or the lethal electical energy pulse,) it is interlocked and caged.

Compressed gases: This is another area where adequate precautions have prevented injuries at our lab. No injuries or damage have occurred here from high pressure cylinders (thousands of psig) although these clearly represent a potential lethal hazard. All intermediate pressures (tens or hundreds of psig) are handled in accord with Section VIII, the Unfired Pressure Vessel Code. In the case of low pressures (fifteen psig or less), especially when combined with cryogenic or vacuum applications, the situation is less clear and our record at NPL includes at least two incidents of which I know. Neither caused injury but significant time, effort, and money were required to replace the two vessels and repair the damaged ones as spares. Both involved liquid nitrogen vessels of the same design specified to operate **essentially** at atmospheric pressure. Each was in the shape of a tall annulus (one meter diameter, one half meter tall, with inner and outer cylinder walls separated by a tenth of a meter) for containing liquid nitrogen. Each was surrounded by our cryostat vacuum when it failed during the initial stage of cooldown from bakeout temperature (below $100°C$) to liquid nitrogen temperature. The method of failure was buckling of the inner cylinder wall inward into the vacuum space followed by weld failure at the top or bottom of the inner wall.

The two collapsed vessels occurred in rather rapid succession and resulted from a combination of causes. The design of the aluminum vessels was inadequate with a working pressure under 21 psia (less than 6 psig), to low to permit aggressive cooldown with liquid nitrogen. The vessels were large and so were not annealed after welding. NPL personnel were not aware of the design pressure limits and believed the tanks safe to at least 15 psig working pressure. (In fact it appears the manufacturer did not calculate the the working pressure until we requested it after the first collapse.) Both collapses also involved poor communications among NPL personnel. The second tank collapse was especially unfortunate and involved poor communication with a foreign visiting scientist whose English was weak. Notice that causes were multiple and the chain of events which led to the accident could have been broken at any one of several points.

Let me emphasize that though the two collapses cost time, effort, and money, there was really little potential for personal injury. Both tanks collapsed inward in nonexplosive manner. In any case, the stainless steel vacuum shell of the cryostats could have safely contained a collapse of a larger tank at higher pressure. As a result of the first collapse, the tank was redesigned and tanks delivered later have a thicker inner wall and working pressure of at least 15 psig. Since roughly half a dozen of the weaker tanks are in use in our cryostats, we have instituted strict liquid nitrogen cooldown procedures and this has prevented further collapses of the these tanks. (We require that any liquid cryogen contained have redundant safety relief devices. In addition a special fill adapter with an orifice and upstream relief valve is now used for cooldown. An extra open vent to atmosphere is also utilized.) Two replacement tanks (of the improved design) were bought and the two collapsed ones rebuilt to the new design as spares.

Lifting devices: The only two accidents with serious potential for lethal injury both involved failure of lifting equipment. Neither caused significant personal injury and only the second caused significant damage. In the first case a four ton shielding block dropped a few inches into its intended position when an eyebolt failed. The bolt hole in the block was shallower than the bolt length so the bolt shoulder did not seat against the block as intended. Unfortunately the bolt experience a modest side load, well within the permissible range if properly seated. Though an NPL employee experienced at rigging was supervising the lift, he failed to notice the gap beneath the shoulder. Luckily the block fell into place rather than crushing valuable equipment.

The second accident was substantially more costly in time and effort even though no significant injury occurred. A loaded cryostat was being lifted from its position in the linac to the cryostat transport dolly when a commercially available hydraulic pulling link (intended to ease initial pickup and final setdown) failed. This hydraulic puller was located between our five ton crane hook and the special strongback we designed and built for our cryostats. The pulling unit was rated at ten tons and so was not retested inhouse for a two ton load. In contrast the crane had been recently serviced and recertified and the strongback had been overdesigned for at least five tons and had been carefully tested with three tons before use on our two ton cryostats. Improper design of the puller for lifting, poor manufacture of a female thread on a shackle, plus incomplete assembly of shackle to bolt all contributed to the failure in which the bolt stripped and slipped from the shackle. Luckily the damage to the cryostat was repairable and the main loss was several weeks time and effort. Since than we have been extremely careful to test ALL links in our lifting rigs.

Summary

I have presented this actual safety history in the hope that others may benefit form our mistakes. I realize some might accuse NPL of carelessness. However,

I believe circumstances at our lab are not much different from those at other labs. Increased awareness of <u>REAL</u> dangers is a prerequisite to a better safety record. Students, technicians, engineers, scientists, and professors must clearly understand that anxiety over possible exposure to penetrating radiation can mask the very real dangers from many other sources. Especially dangerous and unappreciated are lifting accidents with cranes and forklifts. From a broader perspective, of course, the hazards in our nuclear physics labs are generally far less dangerous than the commute to and from work, especially for those of us who cycle to work. I wish you and your lab a safe year.

Discussion D. Will

FRAWLEY: I have a comment about compressor water flow. We had the opposite problem. Our problem was that the water was going through the compressor, heat exchanges too fast. It eroded the heat exchangers and caused failure that contaminated our helium gas with water.

WILL: Which heat exchanger got eroded, the after cooler or the oil cooler?

FRAWLEY: We have had failures in both. The oil cooler failure was a big deal.

WILL: The oil cooler failed to the inside of the system and contaminated oil and helium? And the two after coolers that you had failed to the outside and dribbled all over the place?

FRAWLEY: Right.

WILL: I failed to say that we had to rebraze two of our three after coolers because of water failures to the outside when we received them.

FRAWLEY: We had an engineering consultant come in and look at our heat exchangers. His conclusion was that it was the water velocity that was the problem. So we halved the water velocity.

WILL: In which, in all three of those?

FRAWLEY: At least in the oil cooler phase. That was the only one we were really concerned about.

WILL: Let me just point out that in our case the water first enters the after cooler and cools the helium gas by counterflow. The after cooler is in critical flow, and therefore of some concern to us, it limits the flow through our compressors to 13 gallons a minute, despite the Koch spec of 20 gallons a minute, and that's why our compressors run with a water temperature of 124 degrees Fahrenheit.

That temperature would be significantly less if we had the 20 gallons a minute of flow.

FRAWLEY: We just deliberately reduced it from 20 gallons a minute to about 10.

SMITH: It's about 10. I talked to Pete at Koch in great detail about this, and he said on their tests they run 11 gallons a minute. Our consulting engineer said the oil control heat exchanger is designed for seven and a half gallons a minute. I talked to Pete about the difference in temperatures. Ours is pretty close to what yours are. He said until we get about 145 we won't have any problems at all.

Any damages done by the temperature is not to the oil but to the linings of the motor itself, the insulation.

WILL: I would simply point out to Koch that we found in two places in their literature two different values for the amount of water required and pressure required. Both values are inconsistent with the reported value that I'm just hearing right now of seven to 10 gallons a minute.

In addition, that causes us a secondary problem in that at 10 gallons a minute our plastic piping would be wholly inadequate, and we would have major water leaks all over the place.

The plastic piping is on the deionized side of the heat exchanger in our water cooling system. I would certainly appreciate some sort of definitive statement from Koch to replace the various statements that have been given. I think that would be useful.

LARSON: One of the reasons you use cooling water is to keep the thing cool. Why don't you cool your plumbing with cooling water and bypass the hot source and run a little coolant into your pipes?

WILL: You are saying bypass the compressor and dump some of the cooling water...

LARSON: Put some on the hot pipes.

WILL: That's not necessarily the most desirable thing to do to establish an uncontrolled bypass.

LARSON: I didn't say uncontrolled. I said cool.

PREBLE: Well, you can pretty easily control a bypass situation like that. I guarantee it.

FRAWLEY: Let me just warn you that Koch has not really attempted to convince us this was not our problem. So you should be careful.

WILL: Yes, we will be. We'd like to know what the actual rating of each of the exchangers is and what precautions should be taken on these RS compressors. I think there are quite a few of these out. We have Serial Numbers 49, 50, and 52, I think.

MAN IN AUDIENCE: What instruments do you use to measure your liquid helium levels?:

WILL: We are using AMI superconducting level sensors, 110A. That's the helium sensor, I believe. Model 170 is the liquid nitrogen, also from AMI, and Model 170 is a capacitive probe with oscillator.

MAN IN AUDIENCE: That's for nitrogen?

WILL: Yes, that's for the nitrogen.

MAN IN AUDIENCE: How does the helium one work with the superconducting?

WILL: It's got a little heater at the top and a superconducting wire.

MAN IN AUDIENCE: That's not pressure sensitive to where it gives you any trouble? We have problems in getting those to rebuild.

WILL: If the concern is that sometimes the top of it goes superconducting, and then you get a false flow reading, we have never seen that.

PARDO: It's the other way around. We ran it at triple point.

WILL: I see.

FRAWLEY: Those are pressure dependent.

WILL: Those are pressure dependent problems, and when our pressures are a few pounds over atmosphere.

MAN IN AUDIENCE: What type of silicone diodes are you using to read it out with?

WILL: These are Lakeshore Cryotronics, DT-500s, or something of this sort. The 500 series. They have now replaced it with a new series.

MAN IN AUDIENCE: Is it a button shape or like a transistor can shape?

WILL: We have used the stud mount units by and large. And a few of the others elsewhere.

MAN IN AUDIENCE: Have you had any trouble with the top of the case popping off?

WILL: Yes. We have had several tops of cases pop off, mostly without insult, so far as we can tell.

MAN IN AUDIENCE: They keep working?

WILL: No, what I mean, by without insult is in one case it was concluded that maybe the person wrenched a little tight on it or did something a little harsh on it, but in three or four cases we decided that these tops that popped off just popped off gratuitously.

MAN IN AUDIENCE: We use about 25 per cryostat. They fail at a pretty high rate. Let's see. One other question.

On your vacuum vessels do you have a pressure relief valve?

WILL: Yes there is again a circle seal with "O" ring seal in it. I believe it's a one-inch set to either one or four pounds. In any case it's perfectly adequate for what is a quarter-inch stainless steel vessel.

FRAWLEY: You mentioned de-icing a Beech line. Does that mean you get a buildup of air ice inside?

WILL: I wouldn't say we know exactly what the ice is. Especially as we were installing cryostats. Every so often we would find that we hadn't purged one completely enough. Maybe that wasn't the problem.

In any case, we accumulated some air in the line or some substance that iced. Practically any substance does ice at liquid helium temperatures, and we would get ice in the line.

We would warm up the line by admitting helium gas to our vacuum jacket for a period of a couple hours to assure that the thing was warm. We suspect that you probably could admit it to much less than an atmosphere for a significantly shorter time and still warm the thing up quite adequately in most cases, but in our case, as I said, we took a whole day to do the warmup, and then the pumpout and cool down again took eight hours, because we weren't pressed.

FRAWLEY: You just warmed up and then pump on the plumbing?

WILL: Yes, we warm up, and we either pump on the plumbing, or we use helium gas to force the ice on into the main buffer. We do have good charcoal beds and can get it back out if it will get out of the cold lines and into the main buffer. It's easy to get rid of.

We also have purged the lines to air when we thought we had a very bad blockage.

It should be said that in one case we thought we had a bad blockage, and it proved to be a dropped valve slug that was loose, sitting on the seat there, and not responding to the valve.

PREBLE: Any other questions? If not, the close of the 14th and final session. We thank Ken Chapman and Florida State for having us and being so good to us.

Response to
KPS Model RS Compressor
Cooling System

Koch Process Systems, Inc. (KPS) would like to announce specification changes for the operation of our Model RS Compressor.

Due to recent in-house test data and an independent study completed in conjunction with Florida State University, Koch Process Systems, Inc. was able to identify an error in the cooling system of its Model RS Compressor operation specification. Previously, KPS has instructed its customers to provide up to 20 GPM of water through its water cooling system. For compressors configured with parallel cooling systems between the aftercooler and oil cooler, it is possible to obtain 20 GPM because of the split flow.

Compressors which have the aftercooler and oil cooler configured in series cannot obtain 20 GPM through the system due to pressure drop within the aftercooler. Additionally, recent experience indicates that water flow rates in excess of 12 GPM can adversely affect the heat exchangers because of excessive water tube velocity. In some cases, this has resulted in erosion of the copper tubing.

Studying this effect, Koch Process Systems, Inc. offers its test data results and recommends the following operating conditions be followed in order to avoid any erosion problems and to maintain adequate cooling.

KPS Test Data
Parallel Configuration

	Water Flow GPM	Inlet Water Temp. °F	Compressor Discharge Temp. °F	Aftercooler Helium Temp. °F
Test 1	12.5	70	134	86
Test 2	10	70	141	91
Test 3	8	67	155	88

Water Supply Pressure: 62 psig

Recommended Conditions
Either Configuration

Inlet Water Flow Rate	12 GPM Max.
Inlet Water Temperature	75°F Max.
Compressor Discharge Temperature	170°F Max.
Aftercooler Helium Discharge Temperature	90°F Max.

Any customers with questions in regards to this matter should contact Peter Van Duyne, KPS Standard Cryogenic Systems Product Manager, (617) 366-9111.

BUSINESS SECTION

Mr. McKAY (Yale): I have a few items I would like to dispose of, first a brief report on the budget. Ed Berners not only ran a very nice session for us last year, he turned a profit. So, the SNEAP budget is now approximately $3,000 richer than it's ever been before. This means that the total money in the treasury is about $4,000. Perhaps we should consider just what that money should be used for.

There is a problem in getting out the proceedings and deciding who gets copies. Authors get copies, anyone who has been to the meetings gets copies and after that it gets kind of fuzzy. In the old days when it cost about two or three dollars an incremental copy we sent them out to everyone on the mailing list. This has been whittled down somewhat, but the situation is not clear. I think that it would be the best to say that proceedings are no longer included with the membership. The cost of proceedings if ordered before printing will run something like $12 to $15, so, I am suggesting that we send out a mailing to people and say, if you were not at the meeting and would like a copy of the proceedings, please order them. I don't think it's a real problem at that level of money, but it does mean that those people who would like to get copies will get them. The copies will not be sent to those who would just put them on the shelf.

Mr. LINDER (Washington): This is a question. Could you send one to those that are on the registration list?

Mr. ROWTON (Los Alamos): Maybe what Carl was suggesting is just to send one copy to the lab, if the lab is a member of SNEAP, and then send copies to the attendees of that year's meeting.

Mr. LINDER: That's what I meant.

Mr. McKAY: That is a possibility and the money is in the budget to cover it. If people would like to do that, we certainly could say that membership does include one copy. That is an option and a fairly simple one to administer.

Mr. McNAUGHT (McMaster): One question or problem is that if you allow people to order them, you'd have to have some sort of time limit. You'd surely have to order them when they're being published. You can't go back and get more copies from the publisher.

Mr. BERNERS (Notre Dame): I think I know the answer to that, at least the way our agreement was with World Scientific. We could order books before that agreement was signed in the sense that we could give them a total number of books that we were guaranteeing to buy. If we guaranteed to buy a certain number, in our case 200 books, the cost was $12 a copy. But in the announcement being sent out to libraries, the price is $36 for the book. That means that if you want to provide for ordering, it has to be done before the organizers of that meeting have made their agreement with the publisher. I think that generally means before the meeting actually takes place. The order would have to be received before you come to the meeting.

Mr. McKAY: The mechanism would probably be that we would send out the mailing asking; do you wish to attend? Do you wish to order extra copies or copies?

Mr. BERNERS: Right.

Mr. McKAY: There is a suggestion on the floor that membership include one copy of the minutes. One copy would go to any lab that had paid the $10 membership fee. One copy per $10 fee, in addition to the copies that go out to people who attend the conference and pay the registration to the conference.

Mr. BERNERS: To clear that up, I think it has always been clear that anybody who pays the registration automatically gets a copy.

Mr. McKAY: If there are no other questions or comments, then I would like to ask all those who are in favor of including one copy of the proceedings with each lab membership to please raise your hands.

(Indicate affirmatively)

Mr. McKAY: All those opposed?

(No response)

Mr. McKAY: I declare that the motion passed.

AN UNIDENTIFIED SPEAKER: That was railroaded.

(Laughter)

Mr. McKAY: Has it ever been different in this organization, Charlie?

Often in the past when SNEAP had money in the budget, this was used to help defray the costs of the meeting, especially when sponsored by universities rather than national labs. I would be interested in hearing especially from Ed and Ken about this because they're the ones that are most up to date on the cost and difficulties in running the meeting.

Dr. CHAPMAN (Florida State): As far as we've been able to determine, we're about going to break even. I don't think we have a big problem.

Mr. McKAY: Ed, well, you made a profit.

Mr. BERNERS: The reason we made a profit and the reason we could return money to SNEAP was because after we had set the registration fee, we learned that we were going to get a $5,400 grant from Argonne University's Association Trust Fund to cover the cost of publication. Without that, I think we might have had to go to our Dean for something like one to two thousand dollars for us to break even. He said we could have done that. I think probably the way things have been going with registration at about the hundred dollar level, a lab or institution will be able to break even.

Mr. CARR: (Caltech): What is the cost to run one of these meetings?

Mr. BERNERS: I think total expenses including publication for our meeting were on the order of $16,000.

Mr. McKAY: The next item that I wanted to mention was the problem of providing copies of old minutes. We talked about this at the last meeting and Phil Pepmiller acutally did something about it and looked into the possibility. I have some old minutes sitting on my desk ready to ship to him, but he's gone over to some institute somewhere north of New Haven, It's called Harvard or something.

All I want to do at this point is to report that we started to get it done but it has not yet been accomplished.

Phil was looking into the cost of reproducing some of the old minutes. There are one or two complete sets of minutes now. He was going to reproduce some of the old minutes so they would be available for people to borrow and see what the cost would be if people wanted to get copies.

Mr. LINDGREN (Brookhaven): After last year's meeting I sent out to interested people that wanted copies, the Brookhaven report from '73. It's pretty old, but we still have boxes of those reports. I think Neil Burn mentioned that he had some left over, so people who have had the SNEAP would still have copies around.

Mr. McKAY: I think there are copies for most of the last ten years. There are very few copies of some of the early ones of course since there were very few to start with. A couple people have expressed an interest in getting a complete set. I think that I am willing to do a little bit more to try to trace that down.

Mr. GRAY (Kansas State): Considering the historical aspect and significance of these minutes and the continuing evolution of SNEAP, would it be appropriate to have an archivist for SNEAP to maintain a set of these things in an official capacity?

Mr. McKAY: I think it's not a bad idea. This was the sort of thing that Phil was going to do until, as I say, he wandered off to another institution.

Are you offering?

(Laughter)

Mr. GRAY: I'll do it if nobody else wants to do it, but I need to know somebody that has a complete set of minutes.

Mr. McKAY: Well, I think the first thing is to get a couple extra copies of all minutes made. I think I may contact you then and ask you to take on the job of archivist if you are willing to do so.

Mr. LARSON: Is it possible to consider some other recording medium besides printed paper for these archival copies, I suspect that someone would not want to order up a set and get a shelf full of books, but maybe I'm wrong about that.

Mr. McKAY: There is the possibility of microfilm but I have not found out about the cost of that. I don't really think there is much demand for a full set. It's nice to have one or two of them around, as the old minutes are kind of fun to read.

Mr. LARSON: Preservation for the historians of physics might be the more important.

Dr. CHAPMAN: Isn't it possible that if a full set was placed in at least one of the university libraries, that it would be available to anybody on inter-library loan? We don't claim any copyright on these things so if people care to Xerox all or any part they're interested in, they're welcome to do so.

Mr. ROWTON (Los Alamos): Is that lack of copyright really valid today now tht we're having to publish.

Mr. McKAY: That would not apply to the ones under the new contract, no.

Dr. CHAPMAN: But those copies will be available from the publishers for X-years ahead.

Mr. HYDER (Yale): I would like to inquire of the meeting whether they think it might be an appropriate use of part of the reserve to pay for some fairly detailed subject index of all the minutes. This might possibly be done with the aid of undergraduate students and a PC. One could generate a disk subject index which would mean that some of the remarks that Larry makes year after year after year won't have to be repeated indefinitely.

(Laughter)

Mr. ROWTON: I personally think that would be a good idea. Often I've remembered some remark or some discussion and I'd have to go through three or four volumes until I found it.

Mr. McKAY: Then shall we put that to a vote to see if people are interested in doing that. All those in favor of spending some money looking into the possibility of a set of indices. All those in favor?

(Indicate affirmatively)

Mr. McKAY: Those opposed.

(No response)

Mr. McKAY: I'm not sure who, "we" are in this case, but this is something else that Dick will look into over the next year.

(Laughter)

Mr. McKAY: I will loan him my complete set of minutes.

Are there any topics people would like to bring up? Ed.

Mr. BERNERS: Last year I discovered somewhat to my surprise that I enjoyed the process of editing the minutes of SNEAP-20. When I got the thing all done, I was pleased with the result and I thought I would like to do this again sometime.

This led to an idea that I want to propose now; namely, that anybody else who has ever done editing and finds that he enjoys it, let me know about it because I would like to see if we can't put together a panel of editors who would work every year on at least part of the proceedings and try to give a boost to the people who are arranging the meeting, to take some of the load off their hands.

I think ultimately the ideal way to go would be if all of the authors who bring things in to be published would learn exactly how to do it, and their secretaries would learn exactly how to do it, so that the organizers of the meeting had absolutely no correctional work to do whatsoever on the manuscripts turned in. That part would just go as received. That would take almost all the work off the hands of the organizers in putting together that part of the proceedings. Then if there were a panel of people, half a dozen ought to be able to do it, we could take the discussion transcripts and edit those.

I would think that within something like two months after the meeting is over the publisher would have the completed manuscript. And then four months after the meeting is over the books would be the way out. So, anybody who is as strange as I am and enjoys editing and doesn't mind associating with that kind of weirdo, let me know and we'll see if we can't organize such a panel maybe not for this year, but at least for next year.

Mr. McKAY: For reasons that will become obvious in about two minutes, I am very much in favor of that and I'm sure Ken agrees.

(Laughter)

Mr. McKAY: Having done the minutes before, I would comment that once you've gone through the process it becomes easier. So, to have a group of people that have developed the expertise and would do something on this each year would make it much, much easier. And if you've looked at the copy of the 1986 minutes, I think you'll agree with me that the job that Ed has done on this, is an excellent job.

Are there any other topics that people would like to mention at this point?

(No response)

Mr. McKAY: Then we will go onto the next item, location of next year's meeting. In the last two or three years, we have been going onto a two-year cycle. You'll recall that at last year's meeting, I foolishly suggested that Yale would be more than happy to host the 1988 meeting. Now I've started to realize how much work that's going to be. I told you that we were building a hotel and conference center at Yale specially for SNEAP. It should be finished by 1989.

(Laughter)

Mr. McKAY: We could use the location. How many people have tents?

(Laughter)

Mr. McKAY: We have started to look at alternatives and it looks as though we'll be able to handle the meeting. We've found a hotel where we will be taking over most of the hotel, which I think is useful. There are fewer complaints then from the people in the other rooms.

We've started to look at the excursions. We have had excursions on sailboats, various types of motor boats, buses, double-decker buses etc. I think I have found something else, but I'll leave that as a mystery until next year.

Mr. ROWTON: Where are you going to find that many burros in New Haven?

Mr. McKAY: How did you guess?

At any rate, we're prepared to carry on with our offer. The tentative dates are October 24th to the 27th. I don't believe we need a motion to confirm this. I think we're stuck.

I would like to open the floor for offers for the 1989 meeting.

I have an offer from Oak Ridge and is there anyone else who is going to offer for the '89 meeting?

Mr. KRAUSE (Kansas State): We very much would like to be a host for SNEAP in '89 or '90. We have an EN tandem and we're currently building a booster. If people don't know where Kansas State's at, it's 125 miles west of Kansas City, where Dorothy and Toto live.

(Laughter)

Mr. KRAUSE: It's in the Flint hills of Kansas. Points of interest would be the Eisenhower Museum in Abilene, Kansas. We're right by Fort Riley Military Reservation. There's a considerable amount of history there. So, we want to put our name in the hat.

Mr. McKAY: Very good. And, Norval, would you like to speak for Oak Ridge?

Mr. ZIEGLER (Oak Ridge): Yes. We would certainly like to have SNEAP in 1989. I think we do have some new facilities which were not there when SNEAP last visited. We have a new hotel. We also have a new auditorium. It's part of the ORAU, Oak Ridge Associated Universities. Both of these are of course in Oak Ridge.

The last time SNEAP was there, the meeting were held out at the laboratory which was probably a little inconvenient, not that Oak Ridge has a lot of exciting night life, but probably a little better than the lab.

Mr. JURAS (Oak Ridge): There's a new one.

Mr. ROWTON: Oh, okay.

Mr. ZIEGLER: Plus, Oak Ridge has a massage parlor.

(Laughter)

Mr. ZIEGLER: Well, you know, it's very pleasant in east Tennessee in the fall and I'm sure we could arrange a nice outing, in addition to having good facilities for the meeting.

Mr. ROWTON: Yes, I would be happy to go to either of the two labs that have offered to host it next year or the year after, but it's been a long time since we were in Canada. Can I entice any of the Canadian labs?

Mr. McKAY: Jim, do you have anything to say? Simon Fraser?

Mr. ROWTON: How about Chalk River?

AN UNIDENTIFIED SPEAKER: Neil asked last year for that.

Mr. IMAHORI (Chalk River): I haven't talked to him yet so maybe sometime in a few years time.

Mr. McKAY: I guess we have two firm offers and I think that perhaps if anyone has any questions to direct to either of these two people, we should handle those questions and then put it to a vote.

Are there any questions about either offer?

Mr. EVANS (Los Alamos): What's the location to airports for both these places? Approximate distance to airports.

Mr. KRAUSE: You'd have to fly into Kansas City and then catch Capitol Air, but there is air transportation. We have bus service to the university, too.

Mr. ZIEGLER: It's relatively good air transport to our section, to Knoxville, which is a large airport served by several of the major airlines and there is ground transport between there and Oak Ridge which is only about 30, 40 miles.

Mr. McKAY: If there are no other questions then perhaps, Mr. McNaught and Mr. Lindgren, could you act as scrutineers?

Mr. McKAY: I would ask first of all for those who are in favor of going to Oak Ridge next year and then those for Kansas. So, first of all, raise your hands if you would like to go to Oak Ridge in 1989.

Mr. McNAUGHT: Eleven.

Mr. LINDGREN: Twenty-eight here on this side.

Mr. McKAY: Thirty-nine.

I would ask for votes for Kansas then.

(Pause)

Mr. LINDGREN: I saw seven on this side.

Mr. McNAUGHT: I see 11 on this side again.

Mr. McKAY: Then the meeting will be in Oak Ridge in 1989.

May I suggest that Oak Ridge hold it in Gatlinburg? The fall should be beautiful down there.

Mr. McKAY: I'm sure your suggestion will be taken into consideration. I would like to thank Kansas for their offer and I think the Canadians should also take the subtle hint that was given to get their act going. It is a lot of work to hold this meeting, but it is somewhat fun, and productive in many ways to do it as well. So, I thank the people who are making offers. And keep in mind— Kansas perhaps for the year after Oak Ridge.

The only thing I have left on my agenda is to open the discussion to the floor about the format of meetings, what people have liked about what's happened recently, what changes they would like to see made. This is the time to say what you like about the meetings, what you think were doing right, what you think we're doing wrong.

Dr. CHAPMAN: One question: Do you anticipate the format will be about the same in 1988, three days on the lecture studies of the accelerators and one on boosters or is that still open?

Mr. McKAY: I think it would be that. After this I was going to open up for comments on format as soon as we stick someone with the '89 meeting.

Mr. STARK (McMaster): Since we're passing hints for Canada, I'd like to see Ed's editor business firmed up now.

(Laughter)

Mr. McKAY: Well, I think that perhaps since Ken has things organized for this year, I think it's sort of between him and Ed what they can work out in the sense that I don't think we should impose anything on the people that are running the meeting at this point. If a couple of people can get together, I don't think there's any problem with their offer.

Mr. GRAY I think that the organization of this conference and the pace of the sessions is just about right. There's ample time for discussion and the informality is surviving.

It is refreshing to go to a conference where you're not getting staccato talks ten minutes on center from 8:00 o'clock in the morning until 11:00 o'clock at night. Anything that can be done to preserve that should be done at all costs.

Mr. McKAY: I agree very much. And I think the pacing has worked very well. I know it's very difficult to get that pacing right.

Mr. ROWTON: I think we should compliment Ken very greatly for his intestinal fortitude to set it up this way.

(Applause)

Mr. CARNES (Kansas State): Most people do this anyway, but it would be helpful for those of us who haven't been to very many of these, during status reports and things like that if people would just be sure to make a short statement about the machine configuration. I'm not familiar with a lot of the labs, just to know what accelerator volume and things like that.

Mr. McKAY: I think that's a good point. There is a tendency amongst the old-timers to assume that everybody knows the machines. I think that a brief description of what we're talking about would be very useful.

Mr. WESTERFELDT (T.U.N.L.): I'd like to see some sort of directory maybe in the newsletter. I don't know where all the KN's around the country are or who has FN's and what they're running. I'd like to see some sort of index of machines and insulating gases, tubes upgraded or not, postaccelerator or not, things like that.

Mr. McKAY: There have been surveys prepared, but I do not believe there are any recent ones.

Dr. CHAPMAN: The most recent one was in Berlin, wasn't it, of all the labs that sent any delegates to Berlin? They did a fairly complete survey. Everybody that went was asked to submit the usual two or three page report on the accelerator, ion-source, gases, et cetera, et cetera. These were handed out in a looseleaf binder to all the disciplines.

Mr. HYDER: They are not in the proceedings and there are comparatively few North American machines on that list because the American representation at that European meeting was comparatively small.

Mr. ROWTON: It would be nice if we could get the Oak Ridge compilation updated. It was a useful document and in fact I still have it.

Mr. LEIDICH (Rutgers): A couple years ago when we started the newsletter we got a lot of that data and including insulating gases, charging systems, terminal stabilizers, types of resistors, all that. We started to tabulate it and put it is some format so we could send it out. It's not very complete unfortunately. A lot of labs didn't respond, but I already have a lot of that data. It's on the computer now. I could probably do it if the other labs will fill us in.

If you'll send the form in there's nothing to do. I'll put a newsletter out as soon as I get back and announce it again and have people update, what changes have been made. But a lot of it's already in there. We even calculated the percentage of respondents that used CO_2 and nitrogen versus pure sulphur-hex and things like that.

Mr. McKAY: Perhaps we could get you a copy of the Berlin compilation. I think it would be appropriate if something like that were to be put together that the SNEAP fund would pay for the printing and distribution of it, if people would be agreeable to that.

Mr. LEIDICH: If I could get that Berlin copy I could probably do that pretty quickly.

Mr. McKAY: Dick, could you get a copy for him?

Mr. HYDER: Yes.

Mr. McKAY: Very good. Other comments?

Mr. GRAY   Jerry Dugan might have a rather complete list with the Association of Accelerator Conferences of research facilities involved of all types. He might be approached. He might have that available.

Mr. McKAY: All right. That's another source. We should check that.

This is one of quietest sessions I've seen. I'm not sure whether people are getting old or complacent or are just happy. If there is anything else that people would like to talk about in terms of organization or meetings, now's the time. Are there any other comments?

(No response)

Mr. McKAY: Then as the last part of this Session I would like to make a formal motion of approval and thanks to Ken Chapman and all the Florida State people for what has been so far a very enjoyable meeting.

I ask for a show of hands of all those in favor of such a motion.

(Applause)

Mr. McKAY: Thank you very much Ken.

LABORATORY REPORTS

EN-TANDEM LABORATORY REPORT

Tonny Korsbjerg

Institute of Physics, University of Aarhus

DK-8000 Aarhus C, Denmark

Since very few of you are familiar with the Aarhus facility, I would like, instead of a proper Laboratory Report, to give a brief presentation of our Tandem accelerator.

The present premises were erected in 1972-73, and at the same time, the Tandem accelerator was installed. The accelerator tank with the column structure was inherited from the Niels Bohr Institute, where it had been in operation since 1959. The injector was purchased as a package from DANFYSIK, and also the magnets and steerers were bought commercially, whereas all the beam lines and control electronics were constructed and made at the Institute.

The accelerator is equipped with two injectors, utilizing four different ion sources:

- Duoplasmatron (direct extraction).
- Duoplasmatron (charge exchange).
- Sputtering source.
- Penning source.

Stripping is possible at a number of locations:
- An argon stripper in the terminal.
- A foil stripper in the terminal (150 foils).
- A foil stripper in the HE deadsection (12 foils).
- A foil stripper (30 foils) and a gas stripper in the analyzer image area.

The beam travel ends in one of eight possible target-area beam lines selected by a switch magnet.

The accelerator is operated for an average of ~2800 hours/year. Scheduled maintenance totals ~170 hours/year, and unscheduled maintenance has been kept to ~40 hours/year, all numbers being averages over six years.

No major incidents have taken place during the last many years. We are running on our last HVEC belt. Our lifetime has been ~6500 hours. A replacement belt has been purchased from Vennerlunds, Gothenburg, Sweden. It is a standard conveyor belt, costing a few hundred dollars, but it has proved its usefulness at other facilities long ago.

The LE and the He tube tubes are both the inclined-field type. These tubes have logged more than twelve thousand hours, which is a local record, and they show no signs of ageing so far.

During recent years, target beam line #8 has been established as a test line for the Ultra-High Vacuum system needed for the storage ring that is under construction at the moment.

The obtainable vacuum as well as home-built vacuum-monitoring devices and a complicated bake-out system has been tested here, with satisfactory results.

Other accelerators on the same location are:
- 5-MV Van de Graaff accelerator.
- 2-MV Van de Graaff accelerator.
- 400-kV Van de Graaff accelerator.
- 600-kV Van de Graaff accelerator.
- 300-kV separator.
- 100-kV separator.
- 30-kV separator.

BROOKHAVEN NATIONAL LABORATORY TANDEM FACILITY

R.A. Lindgren

Brookhaven National Laboratory, Upton, New York 11973

Early this year we finished a 7-week period of heavy ion injection into AGS through our 2000 ft. beam line known as HITL (Heavy Ion Transfer Line). We had started out injecting ^{16}O at 105 MeV and during the last month gave them fully stripped ^{28}Si at 186 MeV. This 6.6 MeV/amu beam was accelerated to 14.6 GeV/amu in AGS, and extracted to be used in several relativistic heavy ion experiments. We were supplying 30 µamp pulses of 200 µsec duration, at a repetition rate of 2 seconds.

The vacuum line, pumped mainly by getter strips and activated almost two years ago, is still maintaining the pressure between 10^{-8} and 10^{-10} Torr.

A series of tests was made with 196 MeV pulsed gold beams, supplying 32 µamps of charge 33+, measured at a point close to AGS. We achieved 80% transmission through the transfer line. Slackened 2 µgm foils in the terminal stripped to Q13, with respectable lifetimes. If one allowed the full 80 µamp pulse onto a fresh foil, the foil would immediately disappear. However, if the intensity was decreased to about 10% before striking the foil, then taking less than a minute to come back to full intensity, the lifetime could be several hours. A post stripper foil of 20 µgms was used to strip to +33. These tests were conducted to plan for future AGS booster injection.

Besides AGS injection, there is at present only one other major use of the facility. That is an assortment of space related organizations, all of which are exposing various microchips and integrated circuits directly to heavy ion beams at two- and three-stage energies, at greatly reduced intensities, of course. One aim is to understand more fully single event upsets in solid state devices.

In our two machines we use a total of nine Pelletron chains. Between February 1 and mid-June, we lost four in MP7 and one in MP6. Two had about 25k and three 14k hours. I would like to get some feedback from other Pelletron users on whether or not anyone fastens together pieces from broken chains, to form a full length. Our chains are about 620 links long.

We have been having a very large percentage of 4 and 6 inch idler wheels fail because the urethane rims have become completely separated from the nylon hubs. NEC assures us that as soon as their present supply of pulleys is depleted, they will begin shipping units that have grooves cut in the circumference of the nylon hubs, to give more grip to the molded urethane rims.

The conductive sheave rims that we installed a year ago October were removed in January because we felt that the carbon-loaded nylon was too soft; that is, not resistant enough to abrasion. Our machine was covered with black carbon-loaded nylon dust. We are still using the blue nylon rims with steel contact bands, and oilers. I understand that NEC is now using a new carbon-loaded plastic called Hostalen, an ultra high molecular weight polymer, that has far better abrasion resistance than nylon.

In June we replaced two of the extended tubes in MP7 because they each had a number of cracked insulators and showed up poorly in individual tube conditioning. These were tubes 7 and 8, installed in 1981 and had 28,000 Pelletron hours on them. I will list several interesting observations that were made on them when they were laid out on the floor.

1. All of the glass damage was on the top 135°.

2. All internal damage corresponded with external glass damage.

3. There were many external tracks that had not yet developed enough to cause internal damage.

4. The familiar light grey film covered all around the tubes, 360°; dirty debris was found only on the top 135°.

5. Perhaps 50% of the HVEC installed spark gaps were somewhat loose, and all of these showed a ring of black spark marks on the tube electrode, where the spark elecrode protruded through it.

6. Of these, some of them had spark burns on the sides of the spark electrode, on the retaining rings, and on the tube electrodes.

7. There were many circular marks on the surface of the insulators immediately beneath the actual spark gap, in the middle of the glass, suggesting perhaps the spark gaps should be mounted further from the glass.

Because of the correlation of glass damage on the top side of accelerator tubes and the accumulation of oil scum and other debris in the same area, we wiped down all column and tube glass with damp rags.

The factory installed spark gaps do not make a good enough noninductive connection to the tube electrode. I feel that the clamp-on type used by Dowlish Co. would be superior, and could be placed a bit further from the glass.

During our last run for AGS, tube 8 was causing a lot of tank sparks as we tried to hold 14.5 MV. John Benjamin suggested we open and install some corona points on a rod through a pressure gland in a port in the 7-8 dead section, and adjust them to draw a small amount of current. This would decrease the gradient in tube 8, by increasing the gradient in the rest of the high energy end. We did this, and finished the run with no more disturbance from tube 8. If you ever try this, make sure you have no terminal volts before the points are moved or it might be terminal.

The tubular shielded resistors we made and installed in August 1984 are still holding up. In that time we have had about six fail, either open or change value drastically. The remainder of the 1296 units in MP7 have changed resistance from 800 megohms to between 760 and 780 megohms.

SNEAP Report, September 1987

Robert Carr

Kellogg Radiation Laboratory

California Institute of Technology

Pasadena, CA 91125

ACCELERATOR

The past year was mostly uneventful for our 3UDH-HC Pelletron. We ran the terminal from its rated 3MV down to 150 kV with high reliability, though an attempt to run in tandem mode with 50 kV on terminal didn't produce many protons.

My attempt at mooching a charitable contribution of someone's old switch magnet came to naught, so we sprung for new coils for ours. When we took it apart it was clear that cooling water leakage had been causing an intermittant short from the coils to the shroud and pole piece. It works fine with the new coils.

ION SOURCES

Most of our running this past year was with protons from our RF source in the terminal. The intermittant periods of low source output reported at SNEAP 1986 seem to have been eliminated by the frequently successful take-it-apart-and-put-it-back-together treatment, augmented by replacing the canal and O-rings. One of the viton O-rings had clearly overheated, which may have caused a leak or outgassing that poisoned the discharge.

Recent experiments have used a ^{4}He beam rather than protons, so we just replaced the hydrogen bottle with helium. We are about to install a second metering valve so we can switch between proton and alpha beams without opening the tank.

Our General Ionex 860 sputter source continues to operate reliably. It routinely produces in excess of 40 μA of protons or 100 μA ^{12}C.

ATOMIC ENERGY OF CANADA LIMITED
CHALK RIVER NUCLEAR LABORATORIES

TANDEM ACCELERATOR SUPERCONDUCTING CYCLOTRON (TASCC)
1987 Laboratory Report

Reported by: Y. Imahori

The Chalk River Tandem Accelerator Superconducting Cyclotron consists of a 13 MV MP Tandem accelerator injecting a for sector, K=520 superconducting cyclotron that is designed to accelerate all ions from Lithium to Uranium with an energy resolution of 5 parts in 10 000.

MP TANDEM

Tandem beams produced from September 1986 through August 1987 are shown in Figure 1.

ION	Available (h)	(%)	Typical Parameters (q)	(Mev)	(na)
^{1}H	51	2	1	15	1000
^{12}C	193	7	4	50	1200
^{14}N	10	< 1	5/6	70	400
^{16}O	240	9	6	84	500
^{19}F	76	3	6	123	10
^{28}Si	197	8	9/13	155	10
^{29}Si	156	6	9/12	150	200
^{30}Si	184	7	9/12	150	120
^{32}S	180	7	9/14	167	40
^{34}S	331	13	9/12	140	70
^{35}Cl	20	< 1	8	108	1000
^{37}Cl	234	9	9/13	155	50
^{79}Br	116	4	6	85	300
^{107}Ag	13	< 1	5	51	200
^{115}In	130	5	5	48	200
^{127}I	482	18	6	43	800
^{197}Au	18	< 1	9	100	400

The main points of note are that 18% of the total beam time was provided as I-127 to the cyclotron, and 13% was provided as S-34 for the new 8-PI spectrometer.

A new HICONEX 860 ion source was installed in the ion cage the last week of January this year and has replaced the HICONEX 834.

The previous batch of VIVIRAD resistors installed in the Tandem in April 1986 failed through corona discharge and was removed in October 1986. A new batch of VIVIRAD resistors were installed this

year, first in the high energy end of the Tandem, and later in the low
energy end. Although the new VIVIRAD resistors should allow operation
to 15 MV, operation is currently being limited to 12 MV to avoid the
possibility of damaging the present accelerator tubes. A new set of
eight stainless steel electrode, inclined field, extended high
gradient tubes will be installed in January 1988.

Operation of the Tandem at high voltage has been significantly
improved with the installation of VIVRAD resistors. We have
experienced essentially breakdown-free operation at 12 MV and while an
arbitrary limit of 12 MV has been set, some operation has been carried
out at voltages up to 12.6 MV, again with relatively few breakdowns.

The Tandem will be using 800 Megohm resistors on the new
accelerator tubes and column when upgrading is completed next year.
We have deliberately chosen less resistance than the normal value
(1200 Megohm) in order to maintain sufficient rigidity along the low
energy tubes to avoid the effects of beam loading which causes
uncontrollable RF phase shift associated with injection into the
cyclotron.

The present 800 Megohm VIVIRAD resistors which will be
maintained on the column have logged more than 2000 hours operation.
No replacements have been necessary and no deterioration in resistance
value has occurred, which is routinely checked on each tank opening.
It is felt that the uniform gradient has been solely responsible for
the improved performance of the Tandem.

CYCLOTRON

Following successful runs during the summer of 1986 with
Iodine-127 in the RF 0-mode, the cyclotron was tested on Pi-mode in
the fall.   These tests resulted in unacceptably high temperatures on
the cryostat wall and on the radial diagnostic probes at 50% of full
power, and were postponed.

The next beam used was Bromine-79 accelerated to 20 Mev/u.
During this experiment we encountered continuing midplane vacuum
excursions at the required 60 MHz. These excursions kept tripping the
RF system. During six-month's planned shutdown in December 1986,
repairs and correction on known problems and improvements of
protective and monitoring systems were carried out.

The cyclotron operation resumed June 18, 1987 and 24 hour
operation started in July 23, 1987.

A major reason for this round-the-clock capability is that the
installation of safety interlocks now makes it possible to leave the
system in the hands of operators.   Operators can now operate the
injector, Tandem, Beam line plus the cyclotron's cryogenic system and
diagnostic probes.

FLORIDA STATE UNIVERSITY

TANDEM/LINAC

LABORATORY REPORT   SNEAP XXI   1987

K.R. Chapman, E. Myers, and K.W. Kemper

Tandem

During the last year the tandem has continued to run at terminal potential up to 9.75MV although much of the time has been at 9MV as with the linac this is a standard operating condition.

We have had to change belts twice. The first to replace a belt that had given long and faithful service and the second due to a most unfortunate incident. A screw holding one of the brass pins, that locate the belt guides, sheared. This brass pin lodged between two other belt guides and the corner of the brass base cut through the belt. This left us with two belts of almost identical width throughout almost the entire length of the belt. These two halves then stacked on top of each other and the resultant noise caused the operator to rapidly turn off the drive motor. The motor and drive pulley bearings were replaced in September 1986 and all bearings, motor, drive pulley and alternator, were replaced, largely as a precaution during the second belt change.

We have been troubled on a number of occasions by power outages but this situation is now much improved.

The H.V. (150KV) cable coupling the supply to the ion source failed twice, and we had two fires. The first when the change over switch from regular to emergency power ignited and the second when the grading resistors on the ion source supply cage burned.

In April we installed our double capacity stripper described at SNEAP 1986 and this has proved most useful.

Linac

The super-conducting heavy-ion linac booster (see separate status report) was dedicated on March 20th and began operation (producing nuclear physics data) on June 6th. As of September 18, requests for beams of 95 MeV ^{29}Si, 70 MeV ^{12}C, 96 MeV ^{16}O and 50 MeV ^{6}Li have been satisfied and over 700 hours operation logged with the combined accelerating system.

Polarized Source

The polarized Li source has been used during the past year for a series of calibration and experimental runs for vector polarized ^{6}Li scattering from ^{4}He, ^{12}C and ^{16}O. Previous measurements for ^{6}Li + ^{12}C at 20 MeV taken at Heidelberg and Wisconsin were repeated here and the analyzing power for the three systems agree within experimental error. The on-target beam polarization is 0.5±.04 compared to a maximum theoretical beam polarization of 0.67. Routine beam currents on target are 5-10nA. The chief limitation on the length of time for running the source is the fact that Cs coats the lens surfaces resulting in sparking. In addition Cs consumption of about 3gm/day has limited the running time. A charge exchange canal with a Cs reservoir and limiting apertures for restricting the flow of Cs has been put into operation. The Li beam enters the tandem at an energy of only 20 keV which means that the strong focusing of the low-energy tandem tube dramatically reduces the transmission of the polarized source beam. The low-energy beam optics has been re-studied and the lenses and Faraday cup re-positioned such that the bunching properties of the low-energy buncher are not disturbed for Linac operation. The lenses now provide better matching of the polarized beam to the tandem. The next series of tests with the source will be to determine if the focusing solenoids in the Linac do not depolarize the beam.

University of Iowa

Laboratory Report for SNEAP 1987

Edwin Norbeck

OPERATION

Our CN Van de Graaff with the Poly-C belt has performed well this year. There were no tank openings.

BEAM VIEWING

We have developed a quartz paint that allows any piece of metal to serve the same function as a quartz plate for viewing the beam. We developed this procedure when we had to replace a circular quartz aperture plate in a target chamber. The paint was made by suspending finely powdered quartz in a dilute solution of sodium silicate in water. The copper disk that we used as a substrate was placed on a hot plate at a temperature above 100°C before applying the paint. The water will not evaporate at room temperature. The addition of the sodium silicate adds an orange color to the emitted light. This orange component has a persistence time of a noticeable fraction of a second. The TV image of the beam spot is somewhat brighter than with pure quartz.

LABORATORY PROGRESS REPORT

1987

R.D. KRAUSE, V. NEEDHAM, K. CARNES, M. STOCKLI,

M. WELLS, A. RANKIN, C. HADJISTAMOULOU, and T. TIPPING

JAMES R. MACDONALD LABORATORY

KANSAS STATE UNIVERSITY, MANHATTAN, KANSAS

1. Kansas State University Tandem Van de Graaff Accelerator

The performance of the EN Van de Graaff accelerator at Kansas State University this past year has been very good. During this period, the accelerator has logged 2708.0 hours of beam time operation. Down time during this period totaled 96.0 hours, which required only one tandem opening in March of 1987. This opening was needed to correct an intermittent instability in the tandem voltage control and to perform the usual and routine tandem maintenance.

To date the accelerator has a total of 9333.0 hours of operation on the first inclined field tube that was installed in February of 1984. The second inclined field tube that was installed in January of 1982 has a total of 17457.0 hours of operation. The two high energy inclined field tubes, i.e. tubes 3 and 4 which were installed in November of 1985, have a total of 5178.0 hours of operation. The charging belt, which also was installed in November 1985, has a total of 5234.0 hours of operation.

2. Ion Collision Physics Facility

Considerable progress has been made on the KSU Superconducting Linac booster project. Early in 1987 the Laboratory took possession of an additional 9100 sq. ft. of building space and moved our upgrade activities there. Physical support structures for the Linac have been constructed along with a clean room for cryostat and resonator assembly. The liquid helium (LHe) refrigerator and distribution system have been installed. The Cryogenic Consultants Inc. refrigerator has been tested and provides 600 W of refrigeration with liquid nitrogen (LN2) pre-cooling and 330 W without LN2. Up to 150 liters/hour of LHe can be made with pre-cooling. Final testing of the distribution system is being conducted at this time. The large cryostats that house the Linac have been completed by Combustion Engineering, Inc., and leak-tested. They are now being electropolished and should be delivered to KSU in November. The Niobium split-ring resonators underwent their first round of testing at Argonne National Lab in June. This testing revealed some deficiencies in the high-β resonators: depressed Q-values were

observed in two resonators and mechanical problems prevented testing
of two others. Corrective measures have been taken and testing will
resume in October. Assembly of the Linac is expected to be complete
in mid-1988 and regular operation is scheduled for the following
year. Professor Gray is giving a more complete status report on the
Linac at this meeting.

Computer operations at the J.R. Macdonald Laboratory are also
being affected by the upgrade. After years of faithful service, our
PDP 11/34 computer will soon be converted into a control computer for
the new superconducting Linac using software and interfaces developed
at Argonne's ATLAS facility. The process began this January when we
replaced the PDP CPU, memory, etc. with a NISSHO board that
effectively converted the 11/34 into an 11/44 with 2 MBytes of
memory. The final conversion will take place when we have a new data
acquisition system in place to relieve the PDP of its duties. Along
those lines, we now have in-house two VAXStation II/GPX workstations
and a VAX 8250. We hope to receive a third VAXStation later this
year or early in 1988. The VAXStations will be used for multi-
parameter data acquisition, and the 8250 will be used for data
analysis and general purpose computing associated with the
laboratory. We plan to use acquisition and analysis software that
has already been developed at other laboratories and currently have
LISA, XSYS, and Q in-house. All three packages make use of a Bi-Ra
Microprogrammable Branch Driver to drive a CAMAC parallel branch
highway to which experiments will be interfaced. We hope to have at
least one of these systems running and taking data yet this year.

Substantial progress was achieved on the Cryebis in 1987. Its
previously constructed high voltage platforms were moved into the new
target hall. The previously constructed electron gun and its vacuum
support vessel were cleaned and reinstalled in the final position on
the platform. The test solenoid and a new test collector were
attached. The measured transmission through the test solenoid was
found to be close to unity under appropriate conditions. After
conditioning, the pressure in the electron gun vacuum vessel is in
the lower 10E-10 Torr range and rises to the middle or upper 10E-10
Torr range when the cathode is hot. The small high voltage platform
was installed adjacent to the big platform and carries two
electronics racks and the 2.5 ampere 10 kV power supply. This power
supply and one electronic rack are encased and insulated from the
platform, so that they can be individually energized to over 50 kV
relative to the platform. The platforms are designed for 200 kV
relative to ground, but the lack of the 200 kV power supply limited
the testing to 125 kV so far. Successful long-term tests were
performed with the platforms including the installed deionized water
cooling system without the maximum thermal load. The three single-
phase 200 kV isolation transformers passed in addition a sequential
125 kV test under the full load of 13 kVA each. A commercial fence
was installed surrounding Cryebis. The access door is protected with
a fail-safe mechanism which grounds the platform within a second

after the door is opened. The collector power safety interlock
system was installed and is operational. The 25 kVA collector is
under construction. The drift tube-, the extraction-, and the
diagnostic systems are partially under design and partially under
construction. For protection against catastrophic vacuum failures,
the pneumatic controls of a standard UHV 4-inch gate valve were
modified and it was indicated that the nominal closing time of 1
second can be reduced to 50 milliseconds with a corresponding
reduction in lifetime. The 5 Tesla 1 meter solenoid is being
constructed by Cryogenic Consultants in London and the final tests
are scheduled for October 1987.

The J.R. Macdonald Laboratory is supported by the U.S.
Department of Energy, Division of Chemical Sciences.

Lawrence Livermore National Laboratory *

Multi-user Tandem Laboratory (MTL)

Report to S.N.E.A.P. - 1987

Ivan Proctor and Dale Heikkinen

The Physics Department at Lawrence Livermore National Laboratory (LLNL) is building a new tandem Van de Graaff laboratory. The laboratory is being funded by a coalition of users including several LLNL Divisions, Sandia National Laboratories Livermore and the University of California. Construction funding for the building to house the laboratory was committed in November 1985. Construction of the building was completed in January, 1987 with the tandem pressure vessel sited and roughly aligned. First beam through the tandem to the high energy cup was achieved on July 15, 1987.

A detailed description of the planned laboratory has been submitted to Nuclear Instruments and Measurements [1] for publication, so we will only outline some details here. An artists concept of the laboratory at the time the project began is shown in the attached figure. Not shown in the drawing is a 10 foot wide by 10 foot tall service trench which runs the full length of the building under the tandem and zero degree beamline. Large power supplies, vacuum and other beam support equipment are housed in the tunnel and services rise as bundles through the floor to a beamline. The building is metal beam and siding construction with extensive use of local shielding to reduce radiation levels. We currently have the high energy extension built to the first switching magnet and have one operating beam line and one operating ion source. Other beam lines, ion source stations and the SF6 gas handling system are under construction.

The tandem is the former University of Washington injector FN which we have upgraded with a Pelletron charging system, Dowlish titanium spiral-inclined field tubes and a new column resistor set. We are installing a gas handling system for pure SF6 insulation. The machine was turned on in July with nitrogen and carbon dioxide as the insulating gas, so the terminal has not been above 5 MV. Tandem stabilization is currently via corona stabilization under GVM control with long term drift corrections applied from intercepting slit current signals. We plan to rebuild the terminal assembly in the near future, and will incorporate a floating stripper tube with a fast optical feedback loop at that time.

There are nineteen possible experimental locations in the laboratory each, with the exception of the double 90 degree AMS line, only one bend from the high energy end of the tandem. We expect to have ten beamlines operational in the next year. Some of these are new installations, but most are relocated from the LLNL Cyclograaff or from other laboratories. Four ion source locations are available, two for slow pulsed injection at 30 keV primarily for AMS operation and two at 130 keV which can be used for fast pulsed and bunched operation. We have the usual complement of ion source heads including a tritium sputter source and will add a multiple sample AMS source in the next few months.

All controls for the tandem, beam handling system to an experimental location and the radiation monitoring and control system for personnel access are under a distributed computer control system. Control commands are issued from a Hewlett-Packard 300 series supervisor computer via a fiber optic link to H-P series 300 cluster computers physically located near the equipment to be controlled. The cluster computers are interfaced via CAMAC to the equipment to be read or controlled. Cluster computers report back to the supervisor the values actually set or read. The present operating software is temporary and will soon be replaced by a system to be described in Computers in Physics [2] which will also address in some detail the computer system.

The laboratory is unshielded with a perimeter fence which defines an exclusion zone for personnel safety during periods of operation when we are producing significant levels of radiation within the laboratory. We expect that for about 80 per cent of the operational schedule, people can be present in the laboratory during operation. The radiation monitoring and personnel control system occupies a dedicated cluster computer which monitors and records radiation levels and personnel safety interlocks throughout the laboratory. The radiation level interlocks are set to trip on dose rate and integrated exposure limits. For radiation protection, the computer is backed up by a hard wired system which is also interlocked to tandem and ion source operation.

Useage of the laboratory will be primarily by collaborations among the various participants involved in funding its construction. These include materials characterization on micro and macroscopic scales, AMS jointly with the University of California, AMS for LLNL applications in geochronoloy and radiochemical yield diagnostics, nuclear astrophysics, nuclear spectroscopy and pulsed neutronics studies.

* Work performed under the auspices of the U.S. Department of Energy by the Lawrence Livermore National Laboratory under contract W-7405-Eng-48.

[1] Seventh Tandem Conference, April 6-10,1987, to be published in Nuclear Instruments and Methods.

[2] Tom Moore, private communication.

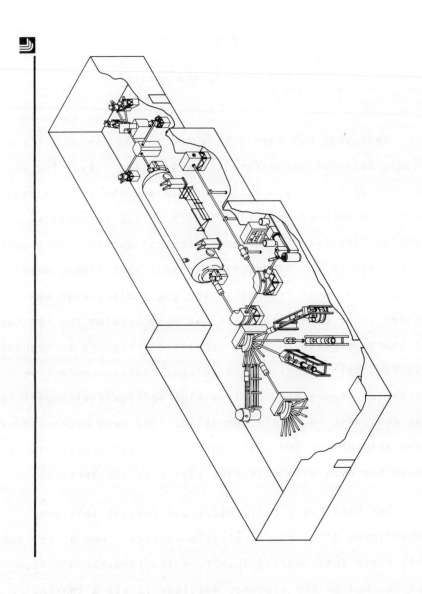

LOS ALAMOS ION BEAM FACILITY LABORATORY REPORT
TO SNEAP XXI 1987

Larry Rowton

Los Alamos National Laboratory
Los Alamos, New Mexico

This year has been a busy and trying period for Los Alamos National Laboratory's Ion Beam Facility. The FN has had essentially a full schedule and the Vertical has been down much of this time. Much of the Vertical's problems involve the telemetry control system. We expect to replace this troublesome equipment next fiscal year. Another source of trouble is the gas cooling heat exchanger. It developed a serious rust problem and the rust of course managed to get liberally distributed throughout the Vertical's terminal and column. Having conductive particles "dance around" in a high voltage environment is not conducive to stable operation. The heat exchanger has been acid etched and passivated. The next problem is to clean the remaining rust particles from the Vertical.

The Vertical's drive motor was rebuilt this year after about 37 years of reliable service. One of the 480 volt phase leads worked lose from its feed-through stud and shorted to the housing. Needless to say a shorted condition like this removes a lot of metal in a short period of time and tends to terminate the operating life of the equipment.

A new slit monitoring system was installed on the Vertical and is working quite well. A new one meter radius of curvature 90 degree magnet and two 12kw power supplies were received from Danfysik this year.

Tank entries for the tandem have been limited to those required for routine foil replacement. Other minor tandem maintenance projects include: rewiring the inflection magnet control ciruit; routine ion source repairs; and a general repair of the door interlock system.

SNEAP Laboratory Report

McMaster University

I.S. Iyer, R. McNaught, J.R. Southon and J.W. Stark

FN

Since our last report at SNEAP '86, the FN accelerator has operated very satisfactorily with only one major shutdown required in August, 1987 to repair a low energy chain break. We have accelerated 11 different ion species, including tritons. Tritons remain as one of our special annual operating features and we would welcome interested users to make use of this rather unique capability.

Our FN is capable of operations at 11 megavolts and this past year we have had over 30 operating days at 10 megavolts or higher.

Our only major problem occurred one weekend in August. The charging system shut down following a spark. Fortunately the researcher on the accelerator detected a noise that was out of the ordinary. Upon investigation, it became apparent that SF_6 was leaking from the H.V. feedthrough at the L.E. base. The accelerator was evacuated in about 7 hours. This leak caused us to lose about 670 pounds of SF_6.

The problem was initiated with a L.E. chain separation. I call it this because it appeared that a rivet had pulled out of the pellet, the aftermath of which broke the aluminum cap off of the H.V. feedthrough at the L.E. end. We have since installed a .5" nylon shield that is standing off about 2" from the feedthrough which we hope will protect us from any future chain break damage in the L.E. base.

Part two of our problem came when we installed our new chain and inspected it to find that in two locations a nylon spacer was missing. This was soon repaired after NEC sent us new master links.

It should be pointed out that had an inexperienced operator or a researcher not familiar with the lab been on duty, they may well have shutdown for the weekend and a significant amount of gas would have been lost.

In May the lab acquired a new NMR of the CERN design and installed it on the analyzing magnet. Our VAX can read the frequency. We set it up on locked mode and, using a DAC and a separate power supply for the trim coil, we can scan through a range of frequencies in 1 kHz steps.

KN

On the KN accelerator at McMaster we are presently in the process of rebuilding the whole terminal deck. A high powered oscillator, the kind that is used at Queen's will be installed. All other power supplies in the terminal will also be replaced with new updated models.

New Ion Sources

The aim of this project is to produce a multi-sample Cs sputter source that will generate higher negative ion outputs than we obtain from our present reflected style sources. The new source is a version of Chapman's inverted sputter source design, but uses a solid spherical ionizer like the latest Middleton source. Sputtered negative ions are focussed by an immersion lens as in the IS3 source developed at Oxford by Nick White. A sample geometry close to that of the inverted source was chosen, as this will allow us to eventually use a rotating multi-sample wheel. A prototype with a single sample holder has been built at a cost of $3000 in material and 300 man hours of machining.

A major advantage of this type of source is its simplicity: the only high voltage required is the -30 kV extraction potential which is also the sputter potential, and all the heaters are at ground. However, the fact that the entire extraction potential is used for sputtering means that the Cs beam heat load on the sputter target is large, and runaway discharges are a problem. Testing to overcome these and innumerable other problems is underway.

University of Notre Dame
Nuclear Structure Laboratory
Notre Dame, Indiana 46556 USA

Report to SNEAP XXI -- 1987

E. D. Berners

Our FN Tandem passed its 19th birthday last July, and the tank cooling coils are showing the effects of old age and Notre Dame water. In April we pulled out all the cooling coils and repaired many leaks. Although the repaired coils passed the bubble test at 200 psi in a water bath, after being put back into the tank they are leaking again. We suspect that flexing the finned tubing during the mounting operation opens up new leaks. We are now looking for replacement coils, and we are also considering smaller coils with blowers.

Belt life is still a problem. Our last belt developed many deep cracks and several large spots with no rubber, after only 4300 hours. We had to remove about 6 inches of width, leaving 72% of the original belt to carry the load. For patching the many deep cracks we tried to find some Kiwi Black Boot Patch (which Ken Chapman mentioned last year), but none could be found. We did find some Kiwi Sport Shoe Patch and applied that to all the cracks. After overnight drying we closed up and commenced running. With only 72% of the belt in place, we ran at 72% of normal tension (1300 pounds instead of 1800 pounds, dynamic). The belt then ran another 870 hours before it disintegrated in flight. By then all the Kiwi patching material had disappeared from the belt and we don't know how long it lasted, or if it did any good at all.

With the pieces of the old belt removed from the tank, we did a car wash on the column, and we are just beginning to run a new belt in a clean machine. With the narrow, trimmed belt running at 1300 pounds tension the pulley bearings ran at 35 °C instead of 50 °C, so we are running our new belt at only 1500 pounds (and 40 °C), to see if the lower temperature will give us longer belt life.

Ohio University SNEAP Lab Report

J.D. Sturbois
Ohio University
Athens, OH 45701
(614) 593-1983

D.E. Carter
Ohio University
Athens, OH 45701
(614) 593-1984

September 25, 1987

1 Charge Screens

As supplied to us by High Voltage Engineering Corporation (HVEC) our machine employed screens made of coarse stainless-steel. The original coarse screens have been replaced by a modification that was developed after initial conversations with Charlie Goldie at HVEC.

1.1 Screen Modifications

The new screens are comprised of two layers of fine screen[1] that are stiffened by a partial layer of coarse screen. This stack is fixed together with spots of epoxy (see figure 1).

The three layers of screen are stacked differently for different points of application. We have added an additional collector screen at the terminal and one on the back of the belt at the base to remove any charge from the back of the belt (see figure 2).

1.2 Tandem Performance with New Screens

For our machine, the voltage holding capability increases as the tandem becomes more stable. The new screen design and placement have increased stability more than a

factor of five. Past average belt life was about 3000 hours with stability degrading after the first 1000 hours of operation. Our most recent belt life was 7822 hours with excellent stability throughout its life.

2 Diode Source Upgrade from 80 kV to 150 kV Extraction

As supplied by HVEC our Diode Injector was designed to extract particles at 80 kV extraction voltage. We have increased the extraction supply to 150 kV and installed a new acceleration tube on the diode source.

2.1 Isolation Transformer Replacement

A major component that was replaced was the 80 kV isolation transformer that is used to supply 208 volt 3 ϕ AC to the welder and associated componets that float at the extraction supply voltage. Cost of a 150 kV isolation transformer was in excess of $10,000.00.

A surplus motor and generator were available at almost no cost (only the shipping charges). This made the motor generator

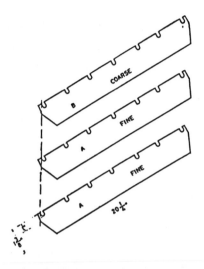

Figure 1: New composite screen design.

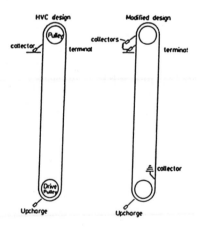

Figure 2: Location of screens

approach much less expensive. This is why the motor generator approach was chosen by OUAL.

2.2 Motor-Generator Design

The set contains a 40 HP motor driving a 25 kVA 208 V 3 ϕ AC generator.[2] The motor and the generator are directly connected by a 36 inch, 6 inch diameter "canvas" fiberglass thick-walled insulating tube.[3] No speed reduction or gearing was employed. The tube was glued[4] to machined steel ends which run on pillow block bearings.[5] The generator is mounted on two sets of insulators[6] separated by an equipotential plane to suppress sparking (see fig. 3).

2.3 Generator Exciter Design

A simple generator exciter (no sensitive solid-state devices) was designed to supply nearly constant excitation current to the generator field winding (fig. 4). Since the generator would not self-excite, a 6 volt lantern battery was used to kick the generator to self-excite. Nearly constant current excitation was achieved using four 60 watt 120 volt light bulbs because of their non-linear resistance characteristic.

2.4 150 kV Extractor Supply

The 150 kV extractor supply was obtained from Del Electronics Corporation[7] at a cost of $3,050.00.

2.5 Diode Injector Accelerator Tube Upgrade

The 150 kV diode injector accelerator tube was purchased from Peabody Scientific[8] at a cost of $6,500.00. This was a direct replacement for the original HVEC tube.

2.6 Cabling From Large Target Room to Vault

Space requirements prohibited locating the motor-generator in the same area as the source. This necessitated cabling the 208 Volt 3ϕ from the generator to the source cage through cable capable of withstanding the 150 kV potential to ground. We used RG220U coaxial cable supplied by Times Fibre Communication[9] at a cost of $3.30 per foot. This cable will support 300 kV DC[1].

3 Duoplasmatron Source Upgrade

The Duoplasmitron source at OUAL has been modified slightly. We have designed, built, and installed a directly heated[10] exchange vapor boiler and exchange canal. The boiler reaches it's ambient temperature in a frighteningly short time of $\approx$ 3 hours. We now use sodium and performance is much more reliable. Power is supplied to the boiler heater through a home-made 30 kV 100 Watt isolation transformer that works very well. Cooling was designed for the ends of the exchange area but abandoned as "not worth the trouble"!

4 ^{3}He Recirculator

OUAL has installed and operated a ^{3}He recirculating unit for the duoplasmitron source. This unit came from the University of Colorado. Bruce Peterson of Ohio State University aided in the installation and initial operation.

---

[1] private communication from Bob Michelak of Del Electronics, (914) 699-2000

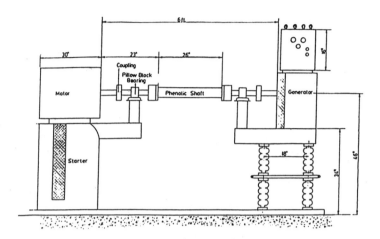

Figure 3: Motor Generator Design

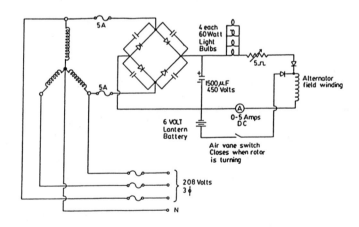

Figure 4: Schematic of Generator Field Exciter

5 Stripper Foil Release Agent

Teepol has always been used as a parting agent at OUAL for our 1-2 $\mu g/cm^2$ machine foils. The passing of time has caused our existing supply of Teepol to "go bad". Extensive testing was done in our quest for a substitute for Teepol. We found that *LTS Lotion Soap*[11] which is a pink handsoap found in the restrooms at OUAL was definitely the best parting agent tested. The best results were obtained using $\approx$ 3 parts soap and 7 parts water.

6 Machine Resistors

After years of testng various types of homemade resistors we are coming to the conclusion that simple designs afford superior performance. We have given up designs that employ epoxy potting. A simple plexiglass tube with carbon resistors in a modified spiral inside has worked very well for us (see fig. 5). The plexiglass tube is ventilated to allow the tank gas to enter and surround the resistors. We have resistors that have been in operation in excess of 5000 hours.

7 Corona Needles

We have developed a method for sharpening our used corona points. The needles are extracted from the base and chucked in a 10,000 RPM "mini" drill press. A stone that is used to dress metal cutting tools is then hand-held to the point to sharpen the needle. Needles are re-inserted into the old base using a drill press collett. The actual time to sharpen each needle is $\approx$ 30 seconds. No performance degradation has been observed compared to new corona points.

8 References

[1] Stainless Steel Wire Cloth, 90 x 90 mesh of 0.005 inch wire, and 50 x 50 mesh of 0.009 inch wire. Newark Wire Cloth Company; 351 Verona Avenue; Newark, NJ 07104

[2] Kato Motor Generator 1800 RPM 25 kVA generator

[3] 6 inch outer diameter canvas tube with 3 inch inner diameter. G5 grade Phenolic Laminate. Resin type is "melamine" and base type is "medium glass cloth" Mil spec. 15037-GMG. The length is 3 feet. Rummel Fibre; 80 Progress St.; Union, NJ 07083; (201) 688-6457

[4] Eccobond 45 black with catylist 15 black. John H. Blaire Co.; Drawer 192; Westerville, OH 43081; (614) 882-6431

[5] Pillow Block Bearings; RPB-203-2, ERPB-203-2

[6] NGK Locke Inc.; 1699 Wall Street; Mount Prospect, IL 60056

[7] Del Electronics Corp.; 250 East Sandford Bvd.; Mount Vernon, NY 10550; (914) 699-2000

[8] Peabody Scientific; P.O. Box 2009; Peabody, MA 01960; (617) 535-0444

[9] Times Fiber Communication; 358 Hall Avenue; Wallingford, CT 06492

[10] GT-3105-20 Fibrox Heating Tape 2 ft. long from Cole Parmer

[11] LTS Lotion Soap; Calgon Corp.; Saint Louis, MO

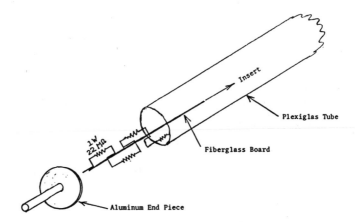

Figure 5: Column Resistor Design

THE ORNL EN TANDEM

P. L. Pepmiller

Oak Ridge National Laboratory
Oak Ridge, Tennessee 37831

The ORNL EN tandem operated reliably this year until late summer when a nagging problem turned into a crippling one. Up until that time, we had been happily carrying on our heavy-ion atomic physics research with the variety of ions available from the Alton series of sources at voltages up to 6 MV. The positron upgrade reported at the Notre Dame SNEAP did not materialize due to a sudden lack of physics justification.

Now for the problem, any help others can provide will be appreciated. We have slightly less resistance in the low energy column than we do in the high energy column. Therefore the column current is slightly higher in the low energy column, up to a point. At voltages of 4 to 5 MV, the situation reverses, that is the high energy column current begins to take off. Over 5 MV, the situation becomes critical, as the upcharge can no longer keep up. This imbalance can reach as high as 125 μA/low, 210 μA/high. It was my contention that the belt must be shedding charge on the first or second active plane since the voltage across a 400 megaohm resistor would be 210 × 400 = 84 kV, which would surely allow us to push an EN tandem to record heights. With this in mind, we shorted 4 planes together, about 10 planes from the HE base plate, and connected an ammeter to them. To our surprise, and my dismay, this meter reads somewhere in between the low and high energy column current meters. This could be explained by the belt losing charge over about the first 20 active planes, while still requiring record voltages across some planes. I have not tried adjusting the drive motor position, but there is little to no evidence for the belt being in physical contact with the belt guides, which by the way are at their maximum spacing. In talks with Zurich and Heidelburg, similiar symptoms have been noted. Heidelburg seems to think that wiping down the column helps, but only temporarily. Help!

414

In other news, which may be reported elsewhere, Zurich tried a Swedish conveyor belt in their machine, but had to remove it due to unacceptable ripple.

A final note is in order here. I have accepted other employment outside of the field of accelerator physics. It has been a pleasure knowing the members of SNEAP and attending the annual meetings. The laboratory is currently deciding who will be my replacement.

This research was sponsored by the U.S. Department of Energy, Division of Chemical Sciences under Contract DE-AC05-84OR21400 with Martin Marietta Energy Systems, Inc.

SNEAP 1987 LABORATORY REPORT
ACCELERATOR LABORATORY, QUEEN'S UNIVERSITY
Kingston, Ontario, Canada   K7L 3N6
by H. Janzen

4 MV VAN de GRAAFF ACCELERATOR

Operation and Maintenance

During the year from 01 September 1986 to 31 August 1987 the Van de Graaff was run for approximately 1900 hours, most of this on experimental runs. This is up somewhat from last year, and reflects the increased use of the machine in PIXE elemental analysis and RBS and channeling work with the new UHV goniometer chamber. Fifty days were spent on machine maintenance, 6 days on test and conditioning, and 22 days on up-grading. Seven tank openings were made during the year for the following reasons:

1) Ion source replaced and belt damage repaired,

2) Tank sparking due to broken shorting rod spring,

3) Belt phase position trigger circuit malfunction,

4) Helium 4 source gas depleted,

5) Terminal voltage instability,

6) Ditto,

7) Ditto, column overhauled and new belt installed.

The primary reason for tank opening #4 was the depletion of helium 4 ion source gas. All source gas reservoirs were replenished, to a pressure of 150 psig, and after this a poor vacuum condition was found to exist in the accelerator tube. On test of the source gas manifold system it was found that two of the thermo-mechanical leaks had developed a cold flow and these were replaced. The ion source, which had accumulated approximately 1700 hours of operation, though still functioning, was also replaced.

The last three tank openings were made, as indicated, in the search for and ultimate correction of a persistant terminal voltage instability that did not condition out, and was of a magnitude such that normal stabilized operation of the machine was unacceptably erratic. Symptoms of the aberration were a short sharp dip in column current of the order of 5 microamps, accompanied by a peaking of the terminal voltage of the order of 0.1 MV, recurring every few seconds. This contravarying relation between column current and terminal voltage suggested a faulty column resistor. The resistors were checked by applying a test voltage successively from plane to plane down the column and measuring the current drawn, first with 10 KV applied, then with 9 V. No bad resistors were found, and the resistance values as calculated from the current drains at both voltages were found to be remarkably consistent, and very close to their rated values of 240 and 400 megohms. The belt was removed to facilitate the cleaning of the column by brushing to loosen accumulated belt dust and then blowing out with compressed air. What appeared to be discharge tracks could be seen on several column and tube glass insulators and these were cleaned using an abrasive particle jet blaster. The old belt, which had accumulated 2869 hours running time, and had considerable damage, was replaced with a new HVEC belt. On the old belt there was evidence of abrasion due to slipping on the inner surface, indicating that it had been run with tension too low, so this was increased from 900 to 1100 lb for the new belt. All belt guides were wiped clean with ethanol and the ceramic insulators were cleaned with Scotch-Brite pads. The column resistors were wiped clean with ethanol and the ends polished by buffing.

On test and conditioning, the new belt was run in very carefully, with tension and tracking maintained by our hydraulic actuators. The terminal voltage was increased slowly, reaching our target value of 4.1 MV in about 12 hours with now no sign of the instability. The new belt is running with low belt ripple but appears to be stretching more than usual, both of which effects may be as a result of the increased belt tension.

Other maintenance included a complete overhaul of the belt charge constant current regulator circuit, suspected of being a possible source of voltage instability, and the NUMAR for the analyzing magnet, which had deteriorated somewhat over the last 15 years of operation.

Upgrading

The Danfysik beam profile monitor system was up-graded by the installation of a two channel multiple head controller which allows the simultaneous display of the beam profile trace from any two of eight monitor heads on two miniature oscilloscopes. In addition to the monitor at the entrance to the analyzing magnet a monitor has been installed between the beam stopper and the switching magnet in the machine room. The simultaneous display of these beam profiles has proved to be very useful for ion beam focussing and steering adjustments to optimize beam transport to the switching magnet and the target room. Other monitors have been installed in the target room as discussed later.

A new auxiliary oil diffusion pump was installed on the analyzing magnet chamber. This chamber is about 2 m from the main vacuum pump station of the Van de Graaff and has a relatively small entrance port so that normal operating vacuum in the chamber was bound to be relatively poor and was suspected of causing focussing difficulty and charge stripping when running high current helium ion beams. The new pump filled with Santovac 5 fluid has a rated pumping speed of 700 L/s and was located on the output side of the analyzing chamber connected by a manifold to three of the chamber output ports for increased pumping speed. With this, the vacuum in the chamber remains acceptably low at about 5 microtorr or less for all operating modes.

The micro-processor based belt charge ripple compensation system[1] has been tested and its effectiveness evaluated by nuclear resonance threshold measurements. The results were encouraging, and on typical runs with and without the belt compensation active the energy spread of the ion beam was reduced to about 750 from about 1250 electron volts. It is expected that even greater reductions will be achieved in the future as more operating experience is acquired in setting the fundamental and higher harmonic correction signal phases and amplitudes. At present a PDP-11/60 assembler is used to load the program into the on-board ROM but this need will be eliminated in future with the installation of an EPROM.

TARGET ROOM

In the target room the only major change has been the set-up of the experiment for the measurement of the $^{12}C(\alpha,\gamma)^{16}O$ reaction cross-section at stellar energies on the #5 (45 deg) beam line. This reaction, involved in the helium burning of stars, has a very small cross-section the precise value of which is of great interest to nuclear astrophysicists[2]. The experiment requires an ultra clean vacuum and ultra pure ^{12}C targets low in ^{13}C isotope. To reduce background the beam line to the target chamber has no apertures and uses two Danfysik beam profile monitors located about 1.5 m apart to allow the operator to focus and steer the ion beam to the target with minimal stray radiation production. High ion beam currents of the order of 50 microamps and 2000 hours of beam time will be required to accumulate meaningful statistics. The experimental set-up is now complete and preliminary tests are in progress.

A remote controller for the beam stopper has been installed in the target room. This is an R.F. link operating at a frequency of 27 Mhz and less than 1 w transmitter power. The transmitter is a battery powered hand held unit that can trigger the insertion or extraction of the beam stopper faraday cup from any location in the target room. A warning horn sounds in the target room whenever the beam stop is extracted to admit beam to the room.

REFERENCES

1) Bate, G.C., M.Sc. Thesis Queen's University (1987).
2) Evans, H.C., et al, Report on Reasearch in Nuclear Physics at Queen's University (1986).

MP-4 N.S.R.L.

UNIVERSITY OF ROCHESTER

C.Cross, M.Culver, M.Koch, R.Roemer, T.Miller

Report to S.N.E.A.P. 87

1. TANDEM OPERATION

The upgraded MP Tandem has performed successfully from the last SNEAP conference to present. We have logged 8039 chain hours and experienced no mentionable problems/failures with the charging system. Our chain pulleys are the NEC conductive rim style, and so far have run well with no signs of excessive wear.

The highest terminal voltage delivered for experimental use was 15.33MV 61.2MEV 7Li 3+. Experimental hours from October 86 to September 87 are 3746 hrs. or 42.8% (Fig.1).

We have, however, experienced multiple D.S.Generator failures after >13.5MV sparks. A total of four failures has led us to better spark damage protection. Initial set up of the generators ultilized standard metal/plastic conduit. After two failures we went to NEC recommended Japanese conduit, still suffering from spark damage we dismissed the high tech conduit for tried and true copper tubing and flair fittings. At this time we installed aluminum shields around the generator at each D.S. We have had no failures since then, but do not have enough operating time above 14MV to claim the problem solved. The repair of the 5/6 stripper and one L.E. power shaft bearing were the only other problems experienced.   Routine maintenance consisted of a foil change in the terminal and 5/6 section, and a seal kit,absorber change for the LE cryopump.

We have moved the 4-5Cu radiation sources from the LE and HE knee to the terminal, we now have three mounted on the tandem. The three sources are spaced equally at the terminal.

Our shorting string system has not performed up to par, we use a grounding strap reel in the terminal retracted by a slo/syn motor via commercial 45 lb. squidding line. We have tried multifilament nylon and monofilament but both results were the same. We are now considering a new position for the system modeled after B.N.L. At our present location within the gradient bars the nylon is sparking to the bars and dead sections breaking, resulting in an unscheduled tank opening. This system and materials worked fine for <12.5MV, but the higher gradients it cannot stand.

2. NEGATIVE ION INJECTOR

We are currently upgrading our negative ion injector in between running experiments. The first step was the installation of a motor/generator rotating power drive shaft system which replaced the original HVEC isolation transformer. Two 40-hp motors are connected to two 25 kV A generators via two 6in-diam Lucite rods. Each generator has an output of 120/208 V ac, three phase, 60 Hz. The generators supply all the power to the injector table. A new 400-KV 1-mA NEC power supply has replaced the original HVEC PRE-ACC. The motor/generator set has been running for about five months now without a problem, however, our NEC PRE-ACC has experienced several electronics related failures. Our old PRE-ACC and ICT are still in working order and will be used in case of emergency.

In late spring we installed a new NEC SNIC source in place of our alpha and off axis sources. The SNIC source power supplies are controlled and monitored by an NEC fiber optic system (Fig. 2). The fiber optic system has worked flawlessly since installed while the SNIC source to date has run unsatisfactorily. Along with the SNIC source we are still running our General Ionex Hiconex 834 source.

The injector vacuum system is presently pumped on by a 16 inch mercury diffusion pump. Work is presently under way to replace this with a 10,000 l/s Varian Cryostack 12 cryopump and a CIT Alcatel 400-l/s turbo pump system. Also under construction is a new corona cage for the injector high voltage tables.

3. GAS SYSTEM

In late May 86' we received 45,450 lbs. of SF6 from Allied Chemical Corporation. This corresponds to an accelerator pressure of 130 PSIA. The old gas mixture of SF6 + N2 + CO2 was kept.

Modifications to the gas handling system for improved drying and purification of the SF6 has begun (see Fig. 3). Additional four gas dryers, each containing 125 lbs. of Alcoa activated alumina (H-152 3/16 pellets) and a 3.5kW. heater, have been installed. Circulation of the gas is now done by a hermetically sealed high pressure gas booster (Spencer Turbine Co. model # GH-10010). This booster pump moves 50 cfm of gas at 120 PSIA. To prevent oil carry over,and cool the gas one of two new water cooled condensers was placed downstream of the booster pump. The second was installed on the output side of the Kinney pump.

Several new stone and cloth filters have been added or existing ones have had their elements replaced. Purification of the SF6 will be aided by the addition of four series 2000 rechargeable gas filter-purifier made by Limco Manufacturing Corporation. These have been purchased but not yet installed. Also built but not installed is a large heat exchanger that will be used in filling the tank.

A Stokes Vacuum blower-roughing pump system (Model #1722-HC) has been installed and tested . The plumbing to the tank is still under construction. It is calculated that with the addition of this pumping system with the present Kinney pump system, evacuation of tandem shall take four hours instead of the present twelve hours.

4. COMPUTER SYSTEM

The data acquisition computer system consists of a D.E.C. VAX 11/750 running the VMS operating which is interfaced with two LSI-11/23 computers running the RT-11 operating system; these LSI-11 computers are interfaced with CAMAC crates. The VAX, and the LSI-11 computers are interfaced through an Able Qniverter, and a Peritek Hex-Q Network Manager. The LSI-11 computers are interfaced to the CAMAC crates via two Jorway Model 411 Branch Drivers. Parallel highway cables connect the branch drivers to CAMAC crates in the control room, and serial highway cables connect the branch drivers to CAMAC crates in the beam line area. Connected to the VAX are two TU77 800/1600 bits per inch tape drives, a 124 Megabyte RM80 disk drive containing the operating system, and a 456 Mb. RA81 disk drive

containing user files. The VAX has 8 Mb. of main memory. The LSI-11
computers each have 256 Kilobytes of main memory along with a Sky
Computer SKYMNK Array Processor, a DLV-11J serial line board, and a
Matrox QRGB-GRAPH color graphics controller.

Fig. 1

TANDEM USE

October 1986 - September 1987

Upgrade	459.5	5.2%
Experimental	3746	42.8%
Machine Conditioning/ Development	2066	23.5%
Scheduled Maintenance	1029	11.8%
Unscheduled Maintenance	273	3.1%
Gas Transfer	837	9.6%
Weekends/Holidays	252	2.8%
Miscellaneous (Startup, Computer Downtime)	97.5	1.1%

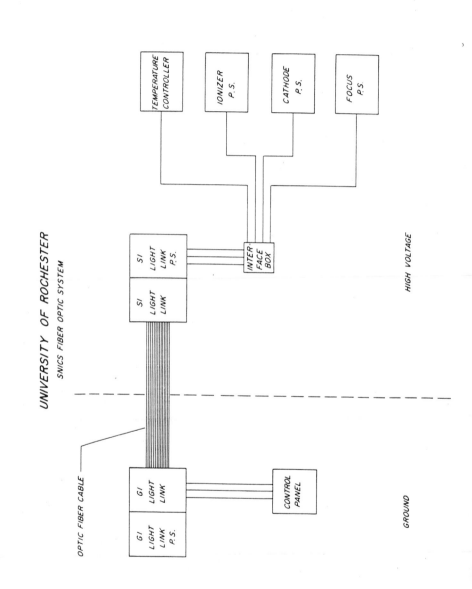

UNIVERSITY OF ROCHESTER

SNICS FIBER OPTIC SYSTEM

425

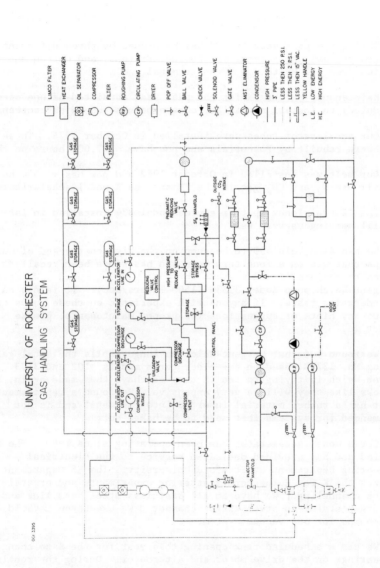

UNIVERSITY OF ROCHESTER
GAS HANDLING SYSTEM

LIMCO FILTER
HEAT EXCHANGER
OIL SEPARATOR
COMPRESSOR
FILTER
ROUGHING PUMP
CIRCULATING PUMP
DRYER
POP OFF VALVE
BALL VALVE
CHECK VALVE
SOLENOID VALVE
GATE VALVE
MIST ELIMINATOR
CONDENSOR
HIGH PRESSURE
3" PIPE
LESS THEN 250 P.S.I.
LESS THEN 2 P.S.I.
LESS THEN 15" VAC.
YELLOW HANDLE
LOW ENERGY
HIGH ENERGY
L.E. LOW ENERGY
H.E. HIGH ENERGY

STANFORD UNIVERSITY

Edward Dillard

During the past year the FN has been used by three different research groups, using beams of Bromine, Protons, and Carbon and energies from 1.7 MV to 7.0 MV on terminal.

Maintenance on the FN for the past year consisted of one opening (scheduled) to install new foils, corona points and adjust screens.

Our accelerator tubes were installed in October 1976. These tubes were rebuilt by Potentials and now have $\approx$ 27,000 hours on them.

Our belt was installed in October 1982 and now has $\approx$ 4500 hours. We still use N_2 and CO_2 gas in the FN and the K model accelerators.

The General Ionex sputter source Model 834 installed in late 1979 is still serving us very well.

We had a Leybold 450 turbo pump on the high energy end of our FN machine that was on a long term loan and has since been recalled. We decided to replace it and upgrade for more pumping speed. We selected a Sargent-Welch pump Model 3133C (1500 liters/sec). It is now on line and undergoing tests. If anyone with experience on these pumps can give us any hints or suggestions for proper maintenance, please contact me.

We found out that Sargent-Welch no longer sells the metal gasket, Catalog No. 1377G, used on the older turbo models 3102 and 3103. Sargent-Welch is trying to see if they can have them made again, but it looks like they will be very expensive. We found a replacement, but it has a rubber material inserted into the metal and is not recommended for systems that need to be baked out.

Our K model accelerator has been operating since 1960. The K was modified and has provided dedicated service to the Electrical Engineering Department at Stanford University. The EE Department utilizes the K for Rutherford scattering experiments and crystal lattice channeling. We have an all stainless steel beam line and a new larger stainless steel target chamber. We use three Leybold 450 turbo pumps for our vacuum system.

We had a scheduled tank opening this year for the K to change out the bearings on the drive motor and alternator. During the routine cleanup and testing, we also decided to change the source.

LABORATORY OPERATIONS REPORT
Tandem/Linac Accelerator

Physics Department
SUNY at Stony Brook
Stony Brook, NY 11794-3800

R. Bundy, M. Cole, A. Gutschow, J. Hasstedt, R. Lefferts,
J.W. Noe, J. Preble, C. Purzynski, J. Sikora, and H. Uto

As discussed in detail in separate papers, two major problems in the tandem injector interrupted accelerator operation for a total of nearly five months in the "SNEAP year" ending October 1, 1987. The first unscheduled downtime started with a Laddertron breakage on December 18th. There were considerable difficulties with the chain re-assembly, and this was finally completed in mid-January only after a new technique was developed to reform the nylon links to their original dimensions.

The more serious problem was the failure of glue bonds in all four tandem tubes during a routine maintenance period in late May. Dowlish Development was able to do complete rebuilds on the four tubes in less than two months, and by August 10th they were once again at Stony Brook. Several additional weeks were spent resolving concerns about the integrity of the glue bonds in the new tubes, re-installing them, and chasing small tank gas leaks. At this writing the experimental program is expected to resume the week of September 28th. Unfortunately, the four months of lost accelerator time came at the busiest season for a University laboratory, and after several summer visitors had finalized their plans. On the positive side, the extended stoppage did allow several major improvement projects to be carried out, as summarized below.

In the remaining seven months of "normal" operation, the linac ran for 3100 hours, bringing the total to date to 13,600. The linac beams requested most often were ^{12}C, ^{16}O, ^{19}F, ^{31}P, and ^{32}S, a change from the strong emphasis on nickel last year. Scheduling has stabilized into a pattern of normally one, or at most two, linac runs per week. The Monday workday is reserved for maintenance, and Monday evenings are available for tandem-only setup beams such as 7 MeV protons (to produce high energy γ-rays) or 25 MeV ytterbium (to simulate fusion products). As always, the system is operated nights and on weekends by the various research groups with the help of comprehensive documentation.

RESISTOR UPDATE

Our experience with resistor protection schemes was summarized at the last SNEAP meeting. Shortly thereafter one further test of "plate" systems was carried out in a short section adjacent to the terminal. Plates were installed at both the top and bottom of the column and offset so as to give staggered plate spacings of approx. 0.5" and 1.5". The resistors were mounted in the 1.5" gaps, where they are exposed to 30% less electric field than in the 1" geometry. Of course, the interplate capacitance, and hence the RF protection, is also reduced by the same fraction.

Resistor values were carefully checked at each tank opening. The resistors soon aged downwards by 8-10%, essentially the same as before. This experience seems to imply that the aging effect is more related to transient over-voltages than high surface electric fields. So far the resistor changes do not include microbreaks detectable at low voltage.

Early this year it was decided to proceed with in-house construction of a full set of BNL-type protection tubes. A resistance of 800 megohms per plane was adopted. This matches the BNL resistors and the nominal value of our remaining yellow RPC resistors and so will allow convenient incremental installation of the new units as they are completed. The first 50 tube assemblies were manufactured in July and August by accelerator staff and summer students, and these are now in place in the high-energy column together with the ten borrowed BNL

units.   In the latter, some very small changes from the   original
800 ± 1 megohms have been detected by digital megging, but in the worst
case the drop is still less than 2%.

The resistor tubes have been mounted into the column using parts
from old HVEC "hangers" for the yellow RPC resistors.   The top ball
and socket joint has a screw, but the bottom joint is still by a
spring contact.   Since the tubes are too large (1 inch dia.) to be
mounted side by side, they alternate left to right on the inside of
the column.   Solid electrical continuity is assured by copper straps
placed across the column.

TANDEM COLUMN IMPROVEMENTS

Also this past summer, a simple brace was installed under each
of the approx. 800 threaded "saddles" that support the lower corners
of the gradient rings.   Previously the saddles tended to rotate to an
incorrect orientation or position during tank openings, and there was
considerable side-to-side play in the gradient hoops.   After this
modification the saddles are quite rigid but can still easily be removed if necessary.

Figure 1 shows the idea. Standard 3/8" dia. copper tubing is cut
with a <u>dull</u> roller-type tubing cutter into 0.600" lengths, and these
are inserted under each saddle. The key point is that the wedge-shaped
ends of the tubing smoothly deform as the saddle is screwed in, and
this allows one or more full turns of adjustment.   A spacer tube of a
harder material or with square ends would not work in the same way.

<u>Figure 1</u>

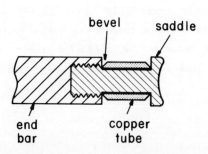

bevel   saddle

end
bar

copper
tube

LINAC VACUUM MANIFOLDS

Up until now the 12 linac modules have been pumped by <u>three</u> small Balzers TPH 270 turbopumps mounted directly (inlet down) on 4.0 inch valves on the top plates. With this system the vacuum levels in the remaining 9 modules are in general rather poor--as high as 10^{-5} torr--because the pumping speed from module to module along the beam line is only a few liters/second. Another shortcoming is the lack of a good turbo interlock, since the gate valve on the module can only be operated manually. (This was a conscious choice to minimize the risk of a <u>really</u> serious accident.)

The maintenance history of these Balzers turbopumps has also been quite unsatisfactory. There have been numerous major failures which necessitated factory repairs costing $1,500 or more, and bearing lifetime is only about six months. In one instance a small puddle of oil was found in a module, after it was warmed up, below the turbo. Another turbopump crashed and left blades lodged in the gate valve! Recently we discovered that some of the pumps repaired by Balzers had been reassembled incorrectly with the foreline at the <u>top</u> of the pump. This can cause contamination of the high vacuum system or degrade the special turbo bearing oil.

Figure 2 shows the new vacuum system designed to address these concerns. The net pumping speed at each module is about 250 ℓ/s, slightly greater than with the turbopump. To minimize backstreaming, the diffusion pump uses polyphenyl ether ("Santovac") and is protected by a water-cooled baffle. The gate valve (from VAT) is interlocked to a Penning gauge in the manifold and a thermocouple gauge in the foreline. All of this costs only about $3,500 per module, compared to $10,000 per module for a comparable turbo system. Further savings are still possible because the diffusion pumps and gauges are readily available used from companies like Duniway Stockroom.

Figure 2

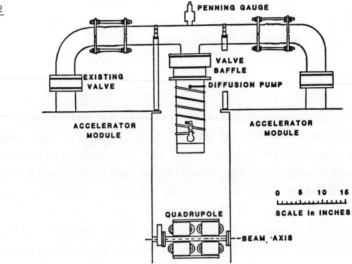

The first of five manifolds under construction has been in operation since August. Vacuum above the pump is less than 1×10^{-7} torr, and cold modules are typically pumped into the mid 10^{-8} torr range. Longterm backstreaming will be evaluated by weighing a collector foil above the pump.

RESONATOR REPLATING

Last year module 8 was replated with a tin/lead alloy, with the result that average fields at 6 watts dissipation increased 25%, from 2.2 MV/m to 2.8 MV/m. This year we have taken advantage of the tandem downtime to replate module 7. Visually satisfactory platings were achieved, but the first cold test of a resonator gave a strikingly poor Q (low 10^7 range) even at low field. We believe that this indicates that the average 0.5 micron thickness used (down from the usual

1.0 micron) must have produced a film that was simply too thin in places. After replating to 1.0 micron the low field Q in one resonator increased to 2.0×10^8. Because of time constraints, the remainder of the tests will be done on-line in October.

CRYOGENICS

The linac cryogenics system continues to operate very smoothly and reliably produces 250 net watts. More recently we have been working on a long helium transfer line for the rebuncher in the target area. Once a week or so the line will be used to top off a 1000 liter cryostat, which in turn will provide 30-liter rebuncher fills every few hours. Boiloff gas will return to the compressor warm. This summer a 6" hole was drilled through a thick concrete wall for the pipe, and thermal studies have been made with Ben-Zvi's program TPF.

QUARTER-WAVE RESONATOR PROJECT

As reported previously the 16 split-loop resonators in the four low-β linac modules are to be replaced with quarter-wave resonators. The entire project is now fully funded and is scheduled for completion by the end of 1988. The cost is quite modest: only $10,000 per resonator including alterations to the cryostats. After the upgrade the practical mass range of the linac will be extended to iodine 127, where we can expect up to 1 pnA at 630 MeV (5.0 MeV/A) in the 27+ charge state. Test results on the prototype QWR were recently reported at the Argonne RF Superconductivity Workshop. The best result so far was 2.5 MV/m at 6 watts, comparable to the high-β split-loops but 50% higher than the present low-β resonators.

TRIANGLE UNIVERSITIES NUCLEAR LABORATORY[*]
REPORT TO SNEAP 1987
Chris Westerfeldt

1. FN TANDEM ACCELERATOR

1.1 General

The TUNL FN tandem accelerator operated for 7107 hours during the past 12 months. Three tank openings were required for stripper foil changes and a charging belt replacement. The charging belt was replaced after a total of 14,392 hours of operation - a laboratory record. After the belt was replaced we had a failure in the drivemotor and it had to be replaced with a spare which we had obtained (used) from HVEC some years ago. We have since obtained a motor on loan from LLNL to backstop us should the present motor also fail.

The tandem operates at terminal potentials between 1.8 and 8.1 MV - limited at the high end primarily by the condition of the accelerator tubes.

1.2 New Intense Polarized Ion Source

The new intense polarized ion source construction project is well underway. All of the major components for the atomic beam apparatus have been completed except for the dissociator and cooled-nozzle assemblies. Completion of the atomic beam apparatus is expected by October 1987. The rf transition units have been designed and prototypes have been constructed and tested. The two sextupole magnets have been constructed and are currently being tested. Initial tests of the polarized atomic beam intensity will begin in October. The ionizer stage design is well underway and actual construction will begin about the end of the year. The original schedule for completing the entire source by the fall of 1988 still seems reasonable.

1.2 New Double-Drift Bunching System

The new double-drift bunching system project described in last years report is nearing completion. All of the hardware is installed in the low-energy injection beamline and most of the new rf control electronics is completed. Final electronics installation and testing will be performed in October.

This buncher is designed to operate at 2.5 and 5.0 MHz for polarized ion beams. For unpolarized beams the bunching frequency is adjustable from 5 MHz down to 39 KHz,
permitting new flexibility in beam pulsing rates for various experiments. The variable buncher tube lengths and buncher frequency will permit a wide range of ions to bebunched. Beam pulse widths of less than 1 ns FWHM are expected for 100 keV injection energies for both polarized and unpolarized protons and deuterons.

----

*Work supported by the U.S. Department of Energy, Office of High Energy and Nuclear Physics, under Contract No. DE-AC05-76ER01067.

1.3   New Beam Transport Power Supplies

We have replaced the power supplies for the magnetic steerers in the laboratory. Twenty new EMI ± 30V, 30A power supplies were installed to handle the steerers and two higher voltage models were purchased to replace the failing power supplies on the high energy quadrupole. Our experience with these power supplies has been very good so far.

We are also currently testing a new high-power op-amp (National LM12) as a replacement for the Darlington pairs now used to drive the magnetic steerer coils. It is hoped that they will be more tolerant of the inductive load and more stable against drift. The prototype has been installed for one month at this time and has had only limited use.

2.   KN ACCELERATOR

2.1   General

The KN Van de Graaff accelerator in the TUNL High Resolution Laboratory is operating routinely again at energies approaching 4.1 MeV. The repairs that we made to the damaged stainless steel electrode accelerator tube last year were very successful. This accelerator tube now has been used for over 8000 hours, mostly at high terminal potentials and the glass insulators are now quite dark from radiation damage. The charging belt installed last year was removed after only 1400 hours of operation when a small tear developed near the center of the belt. The new charging belt is performing very well after approximately 1000 hours of use, exhibiting much less vertical motion than the previous belt.

2.2   Ion Source Modifications

We are constructing a new radio-frequency ion source electronics assembly to replace the original (now 25 years old) electronics presently in use. Major components such as the oscillator assembly (C-XK-TE-18) were purchased from HVEC and only the assembly and wiring of these components will be undertaken by our students. Oscillator related problems account for the majority of our tank openings. We hope that this new oscillator will prove as reliable in our machine as it has in many others. We are also testing a new focus electrode which was constructed using a published design from Utrecht. This focus canal has large pumping slots milled into it to improve the vacuum in the focus gap. We hope that this will reduce the current drain at high focus settings which seem to occur because of a discharge which takes place in the relatively high pressure in this region. Initial tests of this electrode are encouraging.

LABORATORY REPORT TO SNEAP 1987
FOR THE
VICKSI INJECTORS
Hahn-Meitner-Institute, Berlin
reported by
P. Arndt, W. Busse, B. Martin, R. Michaelsen,
W. Pelzer, D. Renner, K. Ziegler

In the past year we have operated VICKSI with two electrostatic injectors using either the 6 MV CN or the 8 UD tandem to inject into the VICKSI cyclotron. Operation with the tandem however has not been as smooth as we had hoped, as different pieces of equipment associated with the tandem went bad forcing us to cut beam periods short or preventing us from delivering beam alltogether.

The operation with the CN injector went pretty much as planned, except that at the beginning of 1987 the beam intensity became more and more instable. This instability was due to the fact that after about 10000 h of operation the "Wennerlund"-CN-belt seems to become mechanically weaker and the movement through the column becomes more and more irregular. The terminal ripple increases until a change of the belt becomes unavoidable. In April 87 a new belt was installed within three days, and at the same time we also fitted yet another modified version of a belt charging screen. Instead of tungsten mesh we now use 0.05 mm sheets of tungsten, resulting in an extremely small abrasion of the belt. Up to now this charging system and the belt have been operating without any problems.

The new capacitive terminal voltage regulation system (CRS) has been included in routine operation after difficulties with the cooling of the high voltage power supply had been solved. A bandwidth of 300 Hz was obtained running the amplifier at a peak to peak voltage of 40 kV into a load of 2000 pF. This new CRS with the final amplifier version has been in operation since January 1987 resulting in a dramatic increase of stability for the VICKSI beams produced with the CN-injector.

Towards the end of last year continuing difficulties with beam transmission through the tandem, inconsistencies between calculated and actual behaviour of the charge state selection system, and a decreasing voltage limit of the offset-quadrupole triplett caused us to decide to check the alignment of the accelerator tubes and of the terminal elements, and to take a closer look at the offset-quadrupole triplett. We found that some of the tube elements were as much as five millimeters out of line. Fortunately we were able to align the tube without dismanteling it. Looking at the offset quadrupole triplett, we found that the offset between the two outer singuletts and the inner singulett was corectly 2.15 mm, however it was mounted in its housing in such a way that the center element was 2.15 mm further away from the beam axis than the outer elements. The correct setting requires, that the outer elements be displaced by 8.35 mm from the beam axis and the inner element 6.20 mm. The voltage limit of the charge selector quadrupole was caused by a relatively thin wire connecting the outer elements and running between the vacuum housing and the oppositely charged center element. At a certain voltage this wire started oscillating until it sparked to the center quadrupole. Replacing this wire by a more rigid metallic tube solved the problem, and we were able to operate the triplett up to ±60 kV without difficulty.

By the end of March this year the tandem was together again and we started tests with low terminal voltages. The transmission for sulphur and nickel beam was better than 80 % of the calculated transmission using tabulated stripping efficiencies. The charge state selection system worked as it should without loss of transmission. The machine was then operated and slowly conditioned up to 8 MV until in early July we were able to run a ^{32}S–beam at 7.9 MV resulting in 960 MeV ^{32}S^{15+} behind the cyclotron, with an intensity of up to 12 nA.

We are confident, that now all elements of the 8 UD are working properly and that we can deliver the beams with the energies and intensities the facility was planned for.

UNIVERSITY OF WASHINGTON NUCLEAR PHYSICS LABORATORY
Seattle, WA 98195

REPORT TO SNEAP 1987

1. Superconducting Booster Linac Construction.

The major part of the Laboratory effort since the last SNEAP meeting has
gone into the completion and initial testing of the superconducting booster LINAC.
We installed the remaining five of the twelve accelerator cryostats and the
remaining buncher cryostat, along with their associated lead–plated copper
resonators, control systems, vacuum systems, and cryogenic distribution system.
We have completed the beam handling systems, including magnets and diagnostics.

We have performed several beam tests. In August we accelerated an ^{16}O
beam to 148 MeV using the entire booster, and transported it into a scattering
chamber for the first time. Our first experiment has been run in September using
a beam of ^7Li at 88.3 MeV, with currents of up to 200 charge nA in the scattering
chamber.

2. Injector Deck

The 300 kV ion source injector deck is now fully operational, with two ion
sources installed, a direct extraction ion source and a Model 860 sputter source.
When the direct extraction ion source is properly tuned for minimum noise, the
voltage noise is 10–15 volts rms. This can become 40–50 volts rms, if the source is
mistuned. The model 860 source has produced analyzed beams of carbon (350 μA),
silver (20 μA), lithium (10 μA), and copper (70 μA) among other species. The 860
source has been extensively modified to reduce beam energy fluctuations and
enhance source and power supply lifetime.

The beam transport system on deck provides mass analysis with resolution
m/Δm $\sim$ 200 using a 90° magnet (r=0.8 m) and object and image slits set at ±3
mm for $\sim$ 90% transmission. The Model 860 is on the 45° port of an additional
switching magnet. An electrostatic quadrupole doublet there insures astigmatism-
free waist–to–waist transport.

The mass analysis system is also used in a high–resolution mode (with slits
set at 0.1 mm) to provide energy resolution E/ΔE $\sim$ 5000 to monitor beam energy
fluctuations, especially important for tuning the DEIS.

The main elevation power supply is a 300 kV, 1 mA Glassman supply
specially modified to provide 5 v p–p noise at 300 kV. The overall deck voltage
noise at 300 kV is 10–15 volts rms. This includes contributions from 20 μA of
corona current (out of 200 μA no–beam load) and 60 Hz harmonics coupled
through the isolation transformer.

With a three–harmonic, single–gap gridded buncher operating at 50 MHz, 50
keV hydrogen has been bunched to 270 ps, 130 keV lithium to 550 ps and 150 keV
oxygen to 650 ps. The bunching efficiency is about 70%.

3. Crossed-Beam Polarized Ion Source.

The crossed beam polarized ion source was installed on the accelerator in May of 1986 and beam from the source was successfully accelerated by the tandem in December although the transmission was and continues to be much lower than expected, only about 50%. The output of the source has been as high as 2 μA. The proton polarization measured in April was 0.86.

The source is now ready to provide beams for experiments which require modest intensity. Work is continuing to improve the source performance, reliability and ease of operation and maintenance. In particular, significant work remains in the development of the Cs beam transport and proton beam extraction and transport. Other specific projects in progress include looking for ways to increase the relatively low output of the atomic beam source, reduce the unexpectedly large unpolarized proton background, improve the control system, and reduce the effects of Cs vapor deposited on electrodes.

4. Tandem Accelerator.

During the past year the tandem ran 3200 hours. A total of 70 operating days were given over to booster installation, and about 54 days were lost due to the belt problem described below.

For several years there have been periods when the tandem terminal potential sagged erratically and the high energy column current increased to as much as 30% higher than the low energy column current. This phenomenon was most common at high terminal voltage. No x rays or increases in beam tube pressure accompanied this phenomenon. In some cases, we observed light flashes along the column in the vicinity of the belt, mostly near the high energy end of the column.

In the past year we entered the tandem three times to fix this problem. The first time, we found a few small bits of metal adhering to the belt guides. Removing these bits did not lead to a permanent fix. The second time we found the belt badly pitted with several holes through it. Also, the belt guides near the upcharge screen were badly damaged. Pieces of belt guide insulator had broken off and there were many spark tracks. The damaged belt guides were replaced, the column was throughly washed and inspected, and the belt replaced.

Within a few weeks, the phenomenon was again noted. We again found the belt guides damaged. However, the new belt showed only minor signs of damage, just a few discolored spots. We replaced all suspect belt guides and carefully inspected the column again. We also replaced the tank gas and encouraged the operators to follow more conservative procedures in raising the terminal voltage. As far as we can tell, the phenomenon has not reappeared; the machine has since run several weeks at terminal voltages in the vicinity of 9.5 MV without problem.

THE WNRE* VAN DE GRAAFF ACCELERATOR
SNEAP REPORT - 1987

by

Sam S.M. Hosein

The performance of our KN 4000 Van de Graaff Accelerator during the past year was adequate. Due to a series of problems, mainly involving the experiments rather than the accelerator itself, we have logged only 560 hours of beam time operation. The machine was operated at a terminal voltage of 4.2 MV in the positive mode to provide a beam for proton irradiation creep experiments.

During this reporting period, four tank openings were made for the following reasons:

1) The beam control selsyn motor, located inside the tank the base of the accelerator, inside the tank, became defective. Further checks revealed that the field winding was open. It was replaced with a spare field coil assembly. In the past we have had similar problems with other selsyn motors.

2) No beam current could be obtained at the first Faraday cup located 2.6 m from the accelerator base. The ion-source plasma would not ignite, even when the RF Oscillator tuning capacitor was adjusted to its full length of travel. The Oscillator tubes (4CX250B) were suspected and, since they could not be checked on our tube tester, they were replaced. The new tubes solved the problem. A point worth noting with regard to the RF Oscillator power supply: we seem to go through a lot of rectifier tubes (6AF3). Most failures are due to arcing and shorting of the elements inside the tubes.

3) To readjust and tighten a loose coupling between the focus control lucite-rod and a variac. The end-play of the rod had increased and this caused the rod to extend out of the key slot of the Focus selsyn motor and became jammed between the motor and the Variac.

4) While using a Veeco vacuum leak detector connected to the beam line close to the base of the accelerator tank, we suspected that there was a small vacuum leak inside the tank. The tank was opened and the accelerator tube was checked using helium gas,

Atomic Energy of Canada Limited
*Whiteshell Nuclear Research Establishment
Pinawa, Manitoba, Canada ROE 1L0

but no leak was found. We were puzzled because we were sure that there was a leak; we closed the tank and some helium gas was injected into it: the leak showed up again. The tank was opened once more and a detailed check was conducted. The leak was found and the cause was a slightly pinched "O" ring between the ion source and the accelerator tube.

Some other beam line equipment problems that we experienced are as follows:

1) A bellows on one of the movable horizontal slit plates of the Image Slit Assembly developed a vacuum leak and had to be replaced.

2) A few power resistors in our General Ionex X-Y beam deflector system overheated and burned the printed circuit board they were mounted on. Both the resistors and the pc board were replaced.

3) The Danfysik beam profile monitor failed to produce the reference markers on the monitor oscilloscope. The fault was traced to the position marker 571 on the beam line. Since the light-emitting diodes in the light forks for the oscillating vane are known to exhibit aging (the light intensity decreases), it became necessary to replace a biasing resistor in the transistor circuitry of the output stage.

The beam instability mentioned in our SNEAP 1986 report was greatly reduced after the following calibrations and modifications:

1) Recalibration of the Image Slit pre-amplifiers, the slit stabilizer control unit and an active attenuator which was installed between the pre-amplifiers and the stabilizer unit. In our system, the attenuator is necessary because the output of the pre-amplifier is much too high to match the input of the stabilizer.

2) Replacement of the original CPU cable, which ran in the same cable tray as the other power cables for the accelerator. The new cable was strung across the room away from any power cables. With the old cable, the CPU input at the stabilizer measured 3 V p-p, with the new cable the input is less than 200 mV.

New developments are under way:

1) A high Energy Dual Magnetic Steerer, together with a tracking unit, is being built in our workshop.

2) The Object, Image and Target Slit Servo Motor Controller, which will be located in the control room, is about 75% completed.

YALE UNIVERSITY
LABORATORY REPORT - S.N.E.A.P. 1987

reported by John McKay

VOLTAGE TESTING

Last year at this time, our ESTU accelerator was almost ready for initial voltage tests. These tests began in January with a standard column, and, after the installation of the Portico, continued until July. We reached a maximum voltage of about 22.3 MV. The tests have shown that the radial K-columns and the Portico work more or less as predicted. Details of the testing program will be detailed in Session III of this meeting.

Flashers:

During the testing, we had difficulty measuring the currents in different parts of the column. A simple circuit was put together, powered by the column current, to flash a neon light. A phototube on the window counts the flashes which can be translated into current.

CHARGING SYSTEM

We have been using the conductive pulley material supplied by N.E.C. without any problems. There was some dust generated, but it precipitates out on the first pulley shields doing no apparent harm. One chain was equipped with old style metal side shims to see if this made any difference in wear. No effect has been seen yet.

In order to control the differential gradient in the inner and outer columns caused by the current drain due to the radiation sources, we are installing a charge pickoff wheel at the portico dead section. A separate power supply will be attached to the inductor of one chain on the low energy end in order to control this current independently.

INSTALLATION PROBLEMS

Power Shaft voltage problem:

There is a slight problem with the Generators in that they put out 140VAC when measured at one atmosphere. It is expected that they will slow down a bit under full SF6 pressure, but this has not yet been checked. At present a Variac is used in the terminal to lower this to 115 volts.

Resistor/Magnet conflict:

A modification was required to the external tube magnets in order to mount the Vivirad resistors. Approximately half an inch had to be machined off each end of the magnet. The operation is not difficult, but care is required to make sure that all cuttings are removed from the magnet.

CURRENT STATUS

At present, the tubes are being installed. A railway has been built to carry the tubes in from the high energy end. This allows the magnets and resistors to be attached outside the tank.

APPENDIX 1

ATTAINABLE ION BEAMS - NEGATIVE SOURCES

This is an informal list of negative ion beams obtained at National
Electrostatics and found in the literature.

It has not been updated for some time. New information and corrections
are always appreciated.

National Electrostatics Corp.
Box 310
Middleton, Wisconsin 53562

Greg Norton
608-831-7600
Fax No. 608-256-4103

Glossary

aea	atomic electron affinity
i.p.	ionization potential
*Na,*K,etc	charge exchange medium

Ion Source References

MID1	Middleton Sputter Source, SNEAP, 1980.
ORSS	Oak Ridge Sputter Source, G. Alton, 1978
AISS	Argonne Inverted Sputter Source, 1981
ANLS	Argonne Sputter Source, SNICS type
NECS	NEC SNICS II, 1986
ANIS	Arhuus Negative Ion Source
UNIS	Middleton Sputtercone Source
HSCS	Hiconex 834 Sputtercone Source, GIC
D+CE	Danfysik 910 & 911 with Charge Exchange
ALPH	NEC Alphatross, 1986
REPG	Radial Extraction Penning Source
SUMM	Ion Source Summary-4th Tandem Conf., 1978

Element	mass	aea	i.p.	matl	gas	ion	I(uA)	ref
H	1.008	0.75	13.6	H/V	--	H-	180.	NECS
H	1.008	0.75	13.6	Ca	NH3	H-	3.	ORSS
H	1.008	0.75	13.6	H/Ti	H2	H-	65.	MID2
H	1.008	0.75	13.6	H/Ti	H2	H-	6.	HSCS
H	1.008	0.75	13.6	H/Ti	--	H-	45.	MID1
H	-----	----	----	-----	----	----	-----	----
D	2.016	--	--	D/Ti	--	D-	35.	MID1
D	2.016	--	--	D/Ti	--	D-	12.	SUMM
D	-----	----	----	-----	----	----	-----	----
He	4.003	<0	24.5	Rb*	He	He-	4.5	ALPH
He	4.003	<0	24.5	Li*	He*	He-	10.	SUMM
He	4.003	<0	24.5	Na*	He*	He-	20.	SUMM
He	-----	----	----	-----	----	----	-----	----
Li	6.939	0.62	5.39	Li	--	Li-	1.	NECS
Li	6.939	0.62	5.39	Na*	Li*	Li-	2.	D+CE
Li	6.939	0.62	5.39	K*	Li*	Li-	2.5	D+CE
Li	6.939	0.62	5.39	Li	02	Li-	1.0	HSCS
Li	6.939	0.62	5.39	LiCu	--	Li-	0.1	MID1
Li	6.939	0.62	5.39	LiCu	--	Li-	2.5	UNIS
Li	6.939	0.62	5.39	Li	--	Li-	1.	ORSS
Li	-----	----	----	-----	----	----	-----	----
Be	9.012	<0	9.32	Na*	Be	Be-	3.4	D+CE
Be	9.012	<0	9.32	CuBe3	Be	Be-	3.4	D+CE
Be	9.012	<0	9.32	CuBe3	02	BeO-	0.9	HSCS
Be	9.012	<0	9.32	Be	02	BeO-	1.8	ORSS
Be	9.012	<0	9.32	BeO	--	BeO-	3.	MID1
Be	9.012	<0	9.32	Be	NH3	BeH-	3.	MID2
Be	-----	----	----	-----	----	----	-----	----
B	10.81	0.28	8.30	Al+B	--	B-	56.	NECS
B	10.81	0.28	8.30	Al+B	--	B2-	73.	NECS
B	10.81	0.28	8.30	Na*	B	B-	2.8	D+CE
B	10.81	0.28	8.30	B	02	BO-	5.	ORSS
B	-----	----	----	-----	----	----	-----	----
C	12.01	1.27	11.3	C	--	C-	90.	NECS
C	12.01	1.27	11.3	C	--	C2-	45.	NECS
C	12.01	1.27	11.3	Na*	C	C-	4.5	D+CE
C	12.01	1.27	11.3	C	--	C-	70.	HSCS
C	12.01	1.27	11.3	C	--	C-	60.	MID1
C	12.01	1.27	11.3	C	--	C2-	20.	ORSS
C	12.01	1.27	11.3	C	CO2	C-	60.	MID2
C	12.01	1.27	11.3	C	--	C-	175.	SUMM
C	-----	----	----	-----	----	----	-----	----
N	14.01	<=0	14.5	Rb*	NH3*	NH2-	4.3	ALPH
N	14.01	<=0	14.5	CTiN	--	CN-	12.	NECS
N	14.01	<=0	14.5	Ti	NH3	NH2-	5.	MID2
N	-----	----	----	-----	----	----	-----	----
O	16.00	1.46	13.6	--	02	O-	550.	NECS
O	16.00	1.46	13.6	Ba	02	O-	63.	ORSS
O	-----	----	----	-----	----	----	-----	----

Element	mass	aea	i.p.	matl	gas	ion	I(uA)	ref
F	19.00	3.40	17.4			F-	89.	NECS
F	19.00	3.40	17.4	Ti	SF6	F-	40.	ORSS
F	-----	----	----	-----	----	----	-----	----
Ne	20.18	<0	21.5					
Ne	-----	----	----	-----	----	----	-----	----
Na	22.99	0.55	5.14					
Na	-----	----	----	-----	----	----	-----	----
Mg	24.31	<0	7.64	Mg	--	MgH2	12.	NECS
Mg	24.31	<0	7.64	Mg	02	MgO-	0.6	HSCS
Mg	24.31	<0	7.64	Mg	02	MgO-	0.6	ORSS
Mg	24.31	<0	7.64	Mg	NH3	MgH-	3.	MID2
Mg	-----	----	----	-----	----	----	-----	----
Al	26.98	0.46	5.98	Al	--	Al-	5.5	NECS
Al	26.98	0.46	5.98	Al	--	Al2-	10.	NECS
Al	26.98	0.46	5.98	Al	--	Al2-	12.	MID1
Al	26.98	0.46	5.98	Al	02	AlO-	2.9	HSCS
Al	26.98	0.46	5.98	Al	02	AlO-	3.	ORSS
Al	-----	----	----	-----	----	----	-----	----
Si	28.09	1.38	8.15	Si	--	Si-	800.	NECS
Si	28.09	1.38	8.15	Si	--	Si-	15.	HSCS
Si	29.09	1.38	8.15	Si	--	Si-	30.	MID1
Si	-----	----	----	-----	----	----	-----	----
P	30.97	0.74	10.5	InP	--	P-	95.	NECS
P	30.97	0.74	10.5	GaP	--	P-	3.	ORSS
P	-----	----	----	-----	----	----	-----	----
S	32.06	2.08	10.4	FeS	--	S-	30.	HSCS
S	32.06	2.08	10.4	Al	CS2	S-	44.	ORSS
S	-----	----	----	-----	----	----	-----	----
Cl	35.45	3.62	13.0	Al	CC14	Cl-	100.	ORSS
Cl	-----	----	----	-----	----	----	-----	----
Ar	39.95	<0	15.8					
Ar	-----	----	----	-----	----	----	-----	----
K	39.10	0.50	4.34					
K	-----	----	----	-----	----	----	-----	----
Ca	40.08	<0	6.11	Ca		CaH3	0.8	ANLS
Ca	40.08	<0	6.11	K*	Ca	Ca-	0.5	D+CE
Ca	40.08	<0	6.11	Ca	02	CaO-	0.2	HSCS
Ca	40.08	<0	6.11	Ca	NH3	CaH-	0.3	ORSS
Ca	40.08	<0	6.11	Ca	NH3	CaH-	3.	MID2
Ca	-----	----	----	-----	----	----	-----	----
Sc	44.96	<0	6.54					
Sc	-----	----	----	-----	----	----	-----	----
Ti	47.90	0.20	6.82	Ti+H		TiH-	22.	ANLS
Ti	47.90	0.20	6.82	Ti+H		TiH-	0.3	HSCS
Ti	47.90	0.20	6.82	Ti+H		TiH-	5.	MID1
Ti	-----	----	----	-----	----	----	-----	----
V	50.94	0.50	6.74	V+H		VH-	40.	ANLS
V	50.94	0.50	6.74	V+H		VH-	0.7	HSCS
V	-----	----	----	-----	----	----	-----	----

Element	mass	aea	i.p.	matl	gas	ion	I(uA)	ref
Cr	52.00	0.66	6.76	Cr	O2	CrO2	0.8	ORSS
Cr	52.00	0.66	6.76	Cr	--	Cr-	0.4	MID1
Cr	-----	----	----	-----	----	----	-----	----
Mn	54.94	<0	7.43	Mn	--	Mn-	0.2	HSCS
Mn	54.94	<0	7.43	Mn	O2	MnO-	1.	ORSS
Mn	-----	----	----	-----	----	----	-----	----
Fe	55.85	0.14	7.87	Fe		Fe-	20.	ANLS
Fe	55.85	0.14	7.87	FeS	--	FeS-	1.6	HSCS
Fe	55.85	0.14	7.87	Fe	O2	FeO-	0.9	ORSS
Fe	55.85	0.14	7.87	Fe	--	Fe-	2.	MID1
Fe	-----	----	----	-----	----	----	-----	----
Co	58.93	0.70	7.86	Co	--	Co-	0.8	HSCS
Co	58.93	0.70	7.86	Co	O2	CoO-	1.3	ORSS
Co	58.93	0.70	7.86	Co	--	Co-	15.	MID1
Co	-----	----	----	-----	----	----	-----	----
Ni	58.71	1.15	7.63	Ni	--	Ni-	125.	NECS
Ni	58.71	1.15	7.63	Ni	--	Ni-	1.3	HSCS
Ni	58.71	1.15	7.63	Ni	O2	NiO-	1.2	ORSS
Ni	58.71	1.15	7.63	Ni	--	Ni-	5.	UNIS
Ni	58.71	1.15	7.63	Ni	--	Ni-	40.	MID1
Ni	-----	----	----	-----	----	----	-----	----
Cu	63.54	1.23	7.72	Cu	--	Cu-	120.	NECS
Cu	63.54	1.23	7.72	Cu	--	Cu-	1.2	HSCS
Cu	63.54	1.23	7.72	Cu	--	Cu-	3.	UNIS
Cu	63.54	1.23	7.72	Cu	--	Cu-	40.	MID1
Cu	-----	----	----	-----	----	----	-----	----
Zn	65.37	<0	9.39	ZnS	--	ZnS-	0.1	HSCS
Zn	65.37	<0	9.39	Zn	O2	ZnO-	2.1	ORSS
Zn	-----	----	----	-----	----	----	-----	----
Ga	69.72	0.30	6.00	GaP	O2	GaO-	1.2	ORSS
Ga	-----	----	----	-----	----	----	-----	----
Ge	72.59	1.20	7.88	Ge	--	Ge-	0.2	HSCS
Ge	72.59	1.20	7.88	Ge	--	Ge-	1.1	UNIS
Ge	72.59	1.20	7.88	Ge	--	Ge-	7.	MID1
Ge	72.59	1.20	7.88	Ge	--	Ge-	1.	ORSS
Ge	-----	----	----	-----	----	----	-----	----
As	74.92	0.80	9.81	GaAs	--	As-	74.	NECS
As	74.92	0.80	9.81	As	--	As-	.5	ORSS
As	-----	----	----	-----	----	----	-----	----
Se	78.96	2.02	9.75	CdSe	--	Se-	15.	ORSS
Se	-----	----	----	-----	----	----	-----	----
Br	79.90	3.36	11.8	Br	--	Br-	33.	SUMM
Br	-----	----	----	-----	----	----	-----	----
Kr	83.80	<0	14.0					
Kr	-----	----	----	-----	----	----	-----	----
Rb	85.47	0.49	4.18					
Rb	-----	----	----	-----	----	----	-----	----
Sr	87.62	<0	5.69	Sr	NH3	SrH-	0.3	ORSS
Sr	-----	----	----	-----	----	----	-----	----

Element	mass	aea	i.p.	matl	gas	ion	I(uA)	ref
Y	88.90	~0	6.38	Y	O2	YO-	1.	ORSS
Y	-----	----	----	----	----	----	-----	----
Zr	91.22	0.50	6.84	Zr	O2	ZrO2	0.7	ORSS
Zr	-----	----	----	----	----	----	-----	----
Nb	92.91	1.00	6.88	Nb		Nb-	7.5	ANLS
Nb	92.91	1.00	6.88	Nb	O2	NbO3	1.	ORSS
Nb	-----	----	----	-----	----	-----	-----	----
Mo	95.94	1.00	7.10	Mo	O2	MoO3	9.2	ORSS
Mo	-----	----	----	-----	----	----	-----	----
Tc	98.00	0.70	7.28					
Tc	-----	----	----	-----	----	-----	-----	----
Ru	101.1	1.10	7.36	Ru	O2	RuO2	1.6	ORSS
Ru	-----	----	----	-----	----	----	-----	----
Rh	102.9	1.20	7.46	Rh	O2	RhO2	0.7	ORSS
Rh	-----	----	----	-----	----	----	-----	----
Pd	106.4	0.60	8.33	Pd	O2	PdO2	2.8	ORSS
Pd	-----	----	----	-----	----	----	-----	----
Ag	107.9	1.30	7.57	Ag+Cs	--	Ag-	3.	ORSS
Ag	107.9	1.30	7.57	Ag	--	Ag-	15.	MID1
Ag	107.9	1.30	7.57	Ag	--	Ag-	3.2	SUMM
Ag	-----	----	----	-----	----	----	-----	----
Cd	112.4	<0	8.99	Cd	O2	CdO-	2.2	ORSS
Cd	-----	----	----	-----	----	----	-----	----
In	114.8	0.30	5.78	In	O2	InO-	2.6	ORSS
In	-----	----	----	-----	----	----	-----	----
Sn	118.7	1.25	7.34	Sn	--	Sn2-	0.1	HSCS
Sn	-----	----	----	-----	----	----	-----	----
Sb	121.8	1.05	8.64	Sb	--	Sb-	16.	NECS
Sb	121.8	1.05	8.64	Sb	--	Sb2-	1.1	ORSS
Sb	-----	----	----	-----	----	----	-----	----
Te	127.6	1.97	9.01	Te	--	Te-	1.2	ORSS
Te	-----	----	----	-----	----	----	-----	----
I	126.9	3.06	10.4	Al	CH3I	I-	25.	ORSS
I	-----	----	----	-----	----	----	-----	----
Xe	131.3	<0	12.1					
Xe	-----	----	----	-----		-----	-----	
Cs	132.9	0.47	3.89					
Cs	-----	----	----	-----		----	----	----
La	138.9	0.50	5.61	La	O2	LaO2	0.4	ORSS
La	-----	----	----	-----	----	----	-----	----
Ce	140.1	--	5.60	Ce	O2	CeO3	0.8	ORSS
Ce	-----	----	----	-----	----	----	-----	----
Tm	168.9	--	5.81	Tm	O2	TmO-	0.6	ORSS
Tm	-----	----	----	-----	----	----	-----	----
Hf	178.5	<0	7.00	Hf	O2	HfO2	2.3	ORSS
Hf	-----	----	----	-----	----	----	-----	----
Ta	180.9	0.60	7.88	Ta	O2	TaO2	3.	ORSS
Ta	-----	----	----	-----	----	----	-----	----

Element	mass	aea	i.p.	matl	gas	ion	I(uA)	ref
W	183.8	0.60	7.98	W	O2	WO3-	6.	ORSS
W	183.8	0.60	7.98	W	--	W-	4.	MID1
W	-----	----	----	-----	----	----	-----	----
Re	186.2	0.15	7.87	Re	O2	ReO2	4.	ORSS
Re	-----	----	----	-----	----	----	-----	----
Os	190.2	1.10	3.50	Os	O2	OsO2	2.	ORSS
Os	-----	----	----	-----	----	----	-----	----
Ir	192.2	1.60	9.00	Ir	--	Ir-	1.	ORSS
Ir	-----	----	----	-----	----	----	-----	----
Pt	195.1	2.13	9.00	Pt	--	Pt-	40.	NECS
Pt	195.1	2.13	9.00	Pt	--	Pt-	3.	ORSS
Pt	195.1	2.13	9.00	Pt	--	Pt-	3.5	UNIS
Pt	195.1	2.13	9.00	Pt	--	Pt-	30.	MID1
Pt	-----	----	----	-----	----	----	-----	----
Au	197.0	2.31	9.22	Au	--	Au-	100.	NECS
Au	197.0	2.31	9.22	Au+Cs	--	Au-	13.	ORSS
Au	197.0	2.31	9.22	Au	--	Au-	10.	UNIS
Au	197.0	2.31	9.22	Au	--	Au-	50.	MID1
Au	-----	----	----	-----	----	----	-----	----
Hg	200.6	<0	10.4					
Hg	-----	----	----	-----	----	----	-----	----
Tl	204.4	0.30	6.11					
Tl	-----	----	----	-----	----	----	-----	----
Pb	207.2	1.10	7.42	Pb	O2	PbO2	0.5	ORSS
Pb	207.2	1.10	7.42	Pb	--	Pb-	<0.1	MID1
Pb	207.2	1.10	7.42	Pb	--	Pb-	0.1	UNIS
Pb	207.2	1.10	7.42	Pb	--	Pb-	0.2	SUMM
Pb	-----	----	----	-----	----	----	-----	----
Bi	209.0	1.10	7.29	Bi	O2	BiO2	1.7	ORSS
Bi	-----	----	----	-----	----	----	-----	----
Po	209.0	1.90	8.43					
Po	-----	----	----	-----	----	----	-----	----
At	210.0	2.80	9.50					
At	-----	----	----	-----	----	----	-----	----
Rn	222.0	<0	10.7					
Rn	-----	----	----	-----	----	----	-----	----
Th	232.0	--	6.95	Th	O2	ThO3	0.8	ORSS
Th	-----	----	----	-----	----	----	-----	----
U	238.0	--	6.08	U	F2	UF5-	0.3	ORSS
U	-----	----	----	-----	----	----	-----	----

APPENDIX II

SNEAP REGISTER
Compendium of Suppliers
Oddball Materials, Goods, and Services

1st Edition, 1987

Edited by
Charles T. Adams
University of Pennsylvania

Contributions for 2nd Edition
now being accepted

Metals - Specialty, Alloys, Braze

Callery Chemical Co.
Callery, PA 10624
(412)538-3510

Cesium metal in ampoules

Fansteel Metals
1 Tantalum Place
North Chicago, IL 60064
(312)689-4900

Tantalum foil, sheet, wire,
rod, bar, tubing

Hamilton Precision Metals
P.O. Box 3014
1780 Rohrerstown Rd.
Lancastrer, PA 17604

(717) 299-2584

ultra-thin Havar foil
windows for gas cells, etc.

Indium Corporation of America
1676 Lincoln Ave.
P.O. Box 269
Utica, NY 13502
(315)797-1630

Indium foil, wire
Joining hi-current elect. joints,
Cryo-pump array surfaces,
"O" Rings

Rembar Co.
69 Main St.
Dobbs Ferry, NY 10522
(914)693-2620

Kovar, Ta, W, Mo

Rhenium Alloys, Inc.
Box 245, 1329 Taylor St.
Elyria, OH 44035
(216)365-7388

Mo/Re Alloy tubing
small percentage Re adds
ductility to Moly.

Spang Specialty Metals
Box 391
Butler, PA 16003
(412)282-3014

Kovar & magnetic materials
smelter

Metals - Contd

Wesgo Division
G.T.E. Products Corp.
477 Harbor Blvd.
Belmont, CA 94002
(415)592-9440

Hi-temp brazing alloys
from 800° to 1770°C

400 Hz - Equipment

Abbott Transistor Labs
Transformer Division
520 Main St.
Fort Lee, NJ 07024
(201)461-4411

400Hz Power transformers

Georator Corp.
9617 Center St.
Manassas, VA 22110
(703)368-2101

400Hz Alternator 1φ, 3 φ
any voltage-current combination
Ideal for terminal operation,
timer belt driven

Metal Screen and Mesh

Buckbee Mears Co.
245 E. 6th St.
St. Paul, MN 55101
(612)228-6374

Electroformed Mesh -
Ni, W, Mo, Au, Cu, S.S.
up to 90% transmission

Unique Wire Weving, Inc.
762 Ramsey Ave.
Hillside, NJ 07205
(201)688-4600

Woven screen
W, Ta, Mo, Pt, Pd, Ti, Cu, S.S.
Tube gridded lens 100x100x.001
81% Txs

Newark Wire Cloth Co.
351 Verona Ave.
Newark NJ 07104

Woven screen
Ni, Pt, etc.

Insulators - AlO_2, BeO_2, Porcelain, Tape

Alberox Corp.
Industrial Park
New Bedford, MA 02745
(617)995-1725

AlO_2 custom precision ground insulators. Sputter source body, cathode insulator

Cole Flex
91 Cabot St.
West Babylon, NY 11704
(516)249-6150

Fiberglass & other wire coverings for magnet wire

Lapp Insulator Co.
Gilbert St.
LeRoy, NY 14482
(716)768-6221

Porcelain station post hi-voltage insulators - building hi-voltage platforms

Permag Magnetics Corp.
2960 South Ave.
Toledo, OH 43609
(419)385-4621

BeO_2 custom made ports.
Heat sink (NBC grade
K-150 Beryllia)

Rim, Inc.
14th Ninth Ave.
Haddon Heights, NJ 08035
(609)547-8041

AlO_2 custon precision ground
Gen. Ion. 860 Source cathode insulator. Dwg. attached

Glass, Quarts, Mica

Asheville-Schoonmaker Mica Co.
910 Jefferson Ave.
Newport News, VA 23607

Mica sheet

Glass,Quartz - Contd.

Esco Optical Products
171 Oak Ridge Rd.
Oak Ridge, NJ 07438
(201)697-3700

Optically flat windows
all grades

Kaufman Glass Co.
P.O. Box 1729
Wilmington, DE 19899
(302)654-9937

Pyrex glass windows
For vacuum chamber portals

National Scientific Co.
219 Paletown Rd., P.O. Box 498
Quakertown, PA 18951
(215)536-2577

Quartz-discs, products

Heaters, Ionizers

Amperex Electronic Co.
230 Duffy Ave.
Hicksville, L.I., NY 11802
(516)931-6200

Inconel Coaxial Heater
900°C Vacuum compatible
.059" - .079" Dia.

Semicon Assoc.
Box 832
Lexington, KY 40501
(606)255-2446

Tungsten cathode dispensers

Spectra-Mat, Inc.
1240 Highway 1
Watsonville, CA 95076
(408)722-4116

Tungsten Ionizers, Dispensers
UNis source ionizers, hi temp
ionizer heaters (1500°C)
ionizers. spherical with potted
 heater

Temptron Engineering, Inc.
7823 Deering
Canoga Park, CA 91304
(818)346-4900

Tantalum Coaxial Heater, Ionizer
1500°C Vacuum compatible
.060" - .100" Dia.

Precision Mechanical Components

PIC
P.O. Box 1004, Benson Rd.
Middlebury, CT 06762
(203)758-8272

Gears, Racks, etc., miniature
bearings, ground rod, Geneva
drives, etc.

Winfred M. Berg, Inc.
499 Ocean Ave.
East Rockaway, NY 11518
(516)599-5010

Same as PIC

Miscellaneous - Materials, Hardware, Services

Betar, Inc.
P.O. Box 896
Summerville, NJ 08876
(201)388-3186

Precision Deep Drilling
Stripper canal

DLDS
1190 Miraloma Way
Suite U
Sunnyvale, CA 94086
(408)737-8111

Cryogenics
100 Kv electrical isolation between
compressor and cold head

Epoxy Technology, Inc.
14 Fortune D
P.O. Box 567
Billerica, MA 01821
(617)667-3805

Optical Coupling Epoxies

General Electric Co.
Chemical Products Plant
1099 Ivanhoe Rd.
Cleveland, OH 44110
(216)266-4329

Electron Tube Emission Carbonate
#117-5-2 Triple Carbonate
for Duo-plasmatron Pt gauze filament

Misc. - Contd.

Lamp Technology, Inc.
141A Central Ave.
Farmingdale, NY 11735
(516)454-6464

Replacement LED's for $TI\frac{3}{4}$ midget
flange lamps (HVEC console)
Yellow 6MF-Y-28P
Green 6MF-G-28P $\approx$ \$3.00 per
Red 6MF-R-28P

Lenape Forge, Inc.
1280 Lenape Rd.
West Chester, PA 19382
(215)793-1500

Delta Gasket
Acc. Tank Door 24" Buna N 300lb Δ Gasket
Recirculator Flange 18" Buna N 300lb Δ Gasket

Multiflex Springwire Corp.
P.O. Box 148-T
Clifton Heights, PA 19018
(215)622-0223

Tube to Column Springs
(72¢ per, 1986)
Dwg. attached

Norco, Inc.
P.O. Box 405
Georgetown, CT 06877
(203)544-8301

Ball Reversing Lead Screw

Safety Lite Corp.
4150-A Berwick Rd.
Bloomsburg, PA 17815
(717)784-4344

Tritiated Titanium

Shell Chemical Co.
One Shell Plaza
Houston, TX 77001
(713)241-6161

Teepol 610 Detergent
Thin film-substrate release agent

458

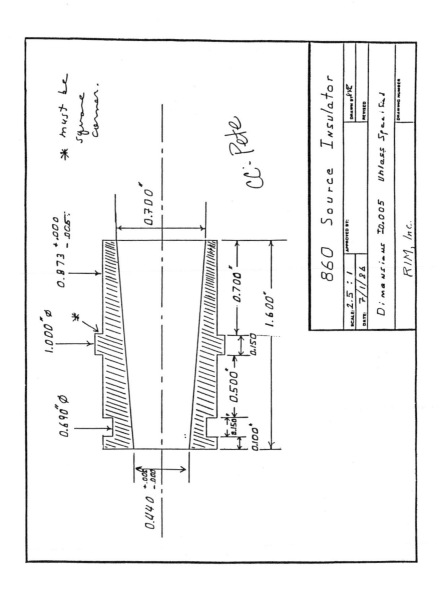

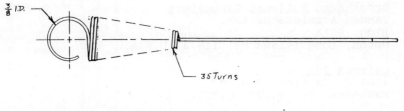

3/8 I.D.

35 Turns

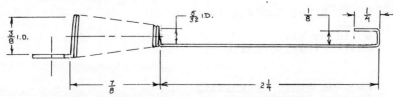

5/32 I.D.

3/8 I.D.

1/8

1/4

7/8

2 1/4

Material: .025 Dia., T-302 S.S. Spring Wire

Quantity: 450 Pieces

CURRENT MEMBER LABORATORIES OF SNEAP

1. State University of New York
    Nuclear Accelerator Lab
    Albany, New York 12222 U.S.A.

2. Argonne National Laboratory
    9700 South Cass Avenue
    Bldg. 203, Physics Div.
    Argonne, IL 60439 U.S.A.

3. Brookhaven National Laboratory
    Tandem Accelerator Lab.
    Bldg. 901A
    Upton, Long Island NY 11973 U.S.A.

4. Caltech 200-36
    1201 E. California Blvd.
    Pasadena, CA 91125 U.S.A.

5. Chalk River Nuclear Lab.
    TASCC
    Building 114
    Chalk River, Ontario
    Canada K0J 1J0

6. Florida State University
    Physics Department
    Tallahassee, Florida 32306 U.S.A.

7. Hahn-Meitner-Institut
    Dep. Physics
    Glienicker Strasse 100
    1000 Berlin 39
    West Germany

8. Kansas State University
    J.R. Macdonald Lab.
    Department of Physics
    Cardwell Hall
    Manhattan, KS 66506 U.S.A.

9. Kentucky, University of
    Department of Physics and Astronomy
    Lexington, KY 40506 U.S.A.

10. Mr. J.D. Larson
    Consultant
    10011 E. 35th St. Terr.
    Independence, MO 64052 U.S.A.

11. **Lawrence Livermore National Laboratory**
    P.O. Box 808   L-397
    Livermore, CA   94550   U.S.A.

12. **Los Alamos Scientific Lab.**
    Ion Beam Materials Laboratory
    MS K764, Group P-10
    Los Alamos, NM   87545   U.S.A.

13. **Los Alamos Scientific Lab.**
    MS K480, P-9
    P.O. Box 1663
    Los Alamos, NM   87545   U.S.A.

14. **New Delhi**
    Prof. G.K. Mehta
    Nuclear Science Centre
    J.N.U. Campus
    P.O. Box 10502
    New Delhi   110067
    India

15. **Notre Dame, University of**
    Department of Physics
    Nuclear Structure Lab
    Notre Dame, Indiana   46556   U.S.A.

16. **Oak Ridge National Laboratory**
    Building 6000
    Oak Ridge, Tennessee   37831-6368   U.S.A.

17. **Oak Ridge National Laboratory**
    Bldg. 5500
    P.O. Box X
    Oak Ridge, TN   37831   U.S.A.

18. Van De Graaff Lab
    Ohio State University
    Physics Department
    Athens, OH   45701   U.S.A.

19. Tandem Accelerator Lab
    Pennsylvania, University of
    Department of Physics
    209 S. 33 Street, E-1
    Philadelphia, PA   19104-3859   U.S.A.

20. Scaife Nuclear Physics Lab.
    Physics Department
    Pittsburgh, University of
    Pittsburgh, PA   15260   U.S.A.

21. <u>Purdue</u> <u>Nuclear</u> <u>Structure</u> <u>Lab.</u>
    Physics Department
    W. Lafayette, IN   47906   U.S.A.

22. <u>Queen's</u> <u>University</u>
    Accelerator Lab.
    Department of Physics
    Sterling Hall
    Kingston, Ontario
    Canada   K7L 3N6

23. <u>Rutgers</u> <u>University</u>
    Nuclear Physics Lab.
    Frelinghuysen Road
    P.O. Box 849
    Piscataway, NJ   08855-0849   U.S.A.

24. Tandem Accelerator Laboratory
    <u>McMaster</u> <u>University</u>
    Rm 105 General Science Bldg.
    Hamilton, Ontario
    L8S 4K1
    Canada

25. <u>Rochester,</u> <u>University</u> <u>of</u>
    Nuclear Structure Lab.
    271 East River Road
    Rochester, NY   14627   U.S.A.

26. Dr. Erle Nelson
    <u>Simon</u> <u>Fraser</u> <u>University</u>
    Dept. of Archaeology
    Burnaby, B.C.
    Canada
    V5A 1S6

27. <u>Stanford</u> <u>University</u>
    Van de Graaff Laboratory
    Physics Department
    Stanford, CA   94305   U.S.A.

28. <u>Stony</u> <u>Brook</u>
    State University of New York
    Nuclear Structure Laboratory
    Department of Physics
    Stony Brook, NY   11794   U.S.A.

29. <u>Strasbourg</u>
    Centre de Recherches Nucleaires
    Services des Accelerateurs
    B.P. 20 / C.R. O
    Strasbourg-Cedex
    67037
    France

30. <u>Triangle</u> <u>Universities</u> <u>Nuclear</u> <u>Laboratory</u>
    Duke University
    Durham, North Carolina 27706 U.S.A.

31. <u>Mr.</u> <u>Don</u> <u>Walker</u>
    Consultant
    9 Country Lane
    Petawawa, Ontario
    Canada
    K8H 3E2

32. <u>Washington,</u> <u>University</u> <u>of</u>
    Nuclear Physics Laboratory
    GL-10
    Seattle, WA 98195 U.S.A.

33. Department of Physics
    <u>Western</u> <u>Michigan</u> <u>University</u>
    Kalamazoo, MI 49008 U.S.A.

34. Atomic Energy of Canada Ltd.
    Van de Graaff Laboratory
    <u>Whiteshell</u> <u>Nuclear</u> <u>Research</u> <u>Establishment</u>
    Pinawa, Manitoba
    Canada
    R0E 1L0

35. Wright Nuclear Structure Lab.
    <u>Yale</u> <u>University</u>
    P.O. Box 6666
    272 Whitney Avenue
    New Haven, Connecticut 06511 U.S.A.

SNEAP 1987 PARTICIPANTS

Charles T. Adams
Physics Department
University of Pennsylvania
209 S. 33 St.
Philadelphia, PA   19104
(215) 898-8832

P. Arndt
Hahn Meiter Institut
Dept. Physics
Glienicher Strasse 100
D-1000 Berlin 39
West Germany

John R. Belden
Glassman High Voltage, Inc.
Rt. 22 Salem Pk B 551
Whitehouse Station, NJ   08889
(201) 534-9007

Edgar D. Berners
Nuclear Structure Lab
Physics Dept.
University of Notre Dame
Notre Dame, IN   46556
(219) 239-7716

Hector Blanchard
Tandem Accelerator Lab
McMaster University
Hamilton, Ontario
CANADA   L87 - 4K
(416) 525-9140   ext. 4046

Tiveron Bruno
I.N.F.N.-Lab. Naz. Legnaro
Via Romea 4
I-35020 Legnaro
ITALY

Tom H. Burritt
Los Alamos National Lab
P.O. Box 1663, MS P234
Los Alamos, NM   87545

Kevin Carnes
Physics Department
Kansas State University
Manhattan, KS   66506
(913) 532-6218

Robert Carr
Kellogg Lab
Caltech 106-38
Pasadena, CA   91125
(818) 356-4583

Don Carter
Accelerator Lab
Ohio University
Athens, Ohio   45701
(614) 593-1984

Federico Cervellera
I.N.F.N.-Lab. Naz. Legnaro
Via Romea 4
I-35020 Legnaro
ITALY

Ken Chapman
Physics Department
Florida State University
Tallahassee, FL   32306
(904) 644-6584

Frank Chmara
Peabody Scientific
P.O. Box 2009
Peabody, MA   01960
(617) 535-0444

Dean Corcoran
Nuclear Physics Lab GL10
University of Washington
Seattle, WA   98195
(206) 543-4085

Samuel L. Craig
Argonne National Lab
9700 S. Cass Ave.
Argonne, IL   60439
(312) 972-4115

Clinton Cross
Univ. of Rochester, NSRL
271 E. River Road
Rochester, NY   14627
(716) 275-4947

Myron Culver
Univ. of Rochester, NSRL
271 E. River Road
Rochester, NY   14627
(716) 257-4947

R. Darling
Serin Physics Lab
Rutgers - The State University
Piscataway, NJ   08854
(201) 932-2404

Edward A. Dillard
Physics Dept.
Stanford University
Stanford, CA   94305
(415) 723-4617

Caleb R. Evans
Los Alamos National Lab
P.O. Box 1663
Los Alamos, NM   87545

Frank Fang
Allied Chemical Corp.
P.O. Box 1139R
Morristown, NJ   07960
(201) 455-4122

L.R. Fell
Dowlish Developments Ltd.
Dowlish Ford Mills
Ilminster
Somerset TA190PF
ENGLAND
4605 2960

Stephen M. Ferguson
Physics Dept.
Western Michigan Univ.
Kalamazoo, MI   49008
(616) 383-4965

John Fox
Physics Department
Florida State University
Tallahassee, FL   32306
(904) 644-6584

Hipolito Gonzalez
Departamento de Fisica
Comision Nacional
  De Energia Atomica
Av. del Libertador 8250
1429 Buenos Aries
ARGENTINA
TX 25073 TANDAR AR

Tom J. Gray
Physics Department
Kansas State University
Manhattan, KS   66506
(913) 532-6029

Chrysostomos Hadjistamoulou
Department of Physics
Kansas State University
Manhattan, KS   66506
(913) 532-6029

Michael Hayes
Kinetic Systems Corp.
11 Mary Knoll Dr.
Lockport, IL   60441

David L. Haynes
Oak Ridge National Lab
P.O. Box X
Oak Ridge, TN   37831
(615) 574-4752

Dale Heikkinen
Lawrence Livermore National Lab
P.O. Box 808 L-397
Livermore, CA   94550
(415) 422-1889

Jean Heugel
Centre de Recherches Nuc.
23, rue du Loess
F-67037 Strasbourg
FRANCE
882 865 64

John R. Hiltbrand
Dept. of Physics
Western Michigan University
1122 Everett Tower
Kalamazoo, MI   49008
(616) 383-4965

Sam Hosein
Whiteshell Nuclear Research Est.
Atomic Energy of Canada Limited
Whiteshell Nuclear Research Est.
Pinawa Manitoba
CANADA R0E 1L0
(204) 753-2311 ext. 2425

Lloyd Hunt
Los Alamos National Lab
P.O. Box 1663, MS K480
Los Alamos, NM   87545

H.R. Mck. Hyder
Wright Nuclear Laboratory
272 Whitney Avenue
P.O. Box 6666
New Haven, CT   06511
(203) 432-3087

Y. Imahori
Chalk River Nuclear Lab
Atomic Energy of Canada Limited
Chalk River, Ontario
CANADA
(613) 584-3311 ext. 2891

Indira S. Iyer
Tandem Accelerator Lab
McMaster University
Hamilton, Ontario
CANADA L85 4K
(416) 525-9140 ext. 4046

Heinz J. Jaensch
Chemistry Dept.
University of California
Santa Barbara, CA   93106

Henry Janzen
Physics Dept.
Queen's University
Stirling 159
Kingston, Ontario
CANADA K7L 3N6
(613) 545-2673

R.C. Juras
Oak Ridge National Lab
P.O. Box X
Building 6000
Oak Ridge, TN   37831-6368
(615) 574-4764

Mark J. Kibilko
Kinetic Systems Corp.
11 Maryknoll Dr.
Lockport, IL   60441

Richard Killian
Allied Signal Corp.
2462 Miller Dr.
Allison Park, PA   15101
(201) 455-4122

Tonny Korsbjerg
Institute of Physics
University of Aarhus
DK-8000 Aarhus C
DENMARK

Bunther Korschinek
Physics Department E17
Technical University
Munich, 8046 Garching
WEST GERMANY
089-3209-2560

Robert D. Krause
Physics Department
Kansas State University
Manhattan, KS   66506
(913) 532-6782

Udo Landmesser
Kali-Chemie Corporation
41 W. Putnam Ave.
Greenwich, CT   06830
(203) 629-7900

S.N. Lane
Oak Ridge National Laboratory
P.O. Box X
Oak Ridge, TN   37831

James Daniel Larson
10011 E. 35th St. Terr.
Independence, MO   64052
(816) 353-1527

Richard R. Leidich
Dept. of Physics
Rutgers University
Box 849 NPL
Piscataway, NJ   08855
(201) 932-2405

Michel Letournel
Centre de Recherches Nuc.
23, rue du Loess BP20/CRO
F-67037 Strasbourg-Cedex
FRANCE
88286564

Carl E. Linder
Nuclear Physics Lab GL-10
University of Washington
Seattle, WA   98195
(206) 543-0609

Robert A. Lingren
Tandem Van de Graaff
Building 901A
Brookhaven National Lab
Upton, Long Island, NY   11973
(516) 282-4581

Rudy Martinez
Los Alamos National Lab
P.O. Box 1663, MS P234
Los Alamos, NM   87545
(505) 667-5211

Karen M. Matejcak
Kinetic Systems Corp.
11 Maryknoll Drive
Lockport, IL   60441
(815) 838-0005

John McKay
Nuclear Structure Lab
Yale University
272 Whitney Avenue
P.O. Box 6666
New Haven, CT   06511
(203) 432-3090

Robert A. McNaught
Tandem Accelerator Lab
McMaster University
Hamilton, Ontario
CANADA   L85 - 4K
(416) 525-9140 ext. 4046

Martha J. Meigs
Oak Ridge National Lab
Bldg. 6000, P.O. Box X
Oak Ridge, TN   37831
(615) 574-4950

Ron Milks
Chalk River Nuclear Lab
Atomic Energy of Canada Limited
Chalk River, Ontario
CANADA   K0J 1J0

Gregory D. Mills
Oak Ridge National Lab
P.O. Box X
Oak Ridge, TN   37831

Edmund G. Myers
Physics Department
Florida State University
Tallahassee, FL   32306
(904) 644-6584

Vincent Needham
James R. Mcdonald Lab
Physics Dept.
Kansas State University
Manhattan, KS   66506
(913) 532-6782

Gregory A. Norton
National Electrostatic
Graber Road Box 310
Middleton, WI 53562
(608) 831-7600

Favaron Paolo
I.N.F.N.-Lab. Naz. Legnaro
Via Romea 4
I-35020 Legnaro
ITALY

Richard Pardo
Argonne National Lab
Bldg. 203
9700 S. Cass Ave.
Argonne, IL   60439
(312) 972-4029

Elizabeth Farrelly Pessoa
Dept. Fisica Nuclear
University of Sao Paulo
C. Postal 20516
01498 Sao Paulo, BRAZIL
TX 1137920 IFSP BR

Arnold N. Peterson
General Ionex Corporation
19 Graff Road
Newburyport, MA   01950
(617) 462-7147

Donald Phillips
Argonne National Lab
9700 S. Cass Ave.
Argonne, IL   60439
(312) 972-4115

Michael Pickett
Koch Process Systems Inc.
20 Walkup Drive
Westborough, MA   01581
(617) 366-9111

Peter F. Potter
ENI, Inc.
100 Highpower Road
Rochester, NY   14623
(716) 427-8300

Joseph Preble
Nuclear Structure Lab
Physics Dept.
SUNY - Stony Brook
Stony Brook, NY   11794
(516) 632-8182

Ivan D. Proctor
Lawrence Livermore Lab   L405
University of California
Livermore, CA   94550
(415) 422-4520

Allan Rankin
Physics Dept.
Kansas State Unviersity
Manhattan, KS   66506
(913) 532-6029

Alan Rice
Caltech Tandem Lab
Caltech 301-38
1201 E. California Bl.
Pasadena, CA   91125
(818) 356-4668

Richard Roemer
Univ. of Rochester, NSRL
217 E. River Rd.
Rochester, NY   14627
(716) 275-2071 (4944)

Larry J. Rowton
Los Alamos National Lab
P.O. Box 1663, MS K480
Los Alamos, NM   87545
(505) 667-5211

Ben Roybal
Los Alamos National Lab
P.O. Box 1663, MS K480
Los Alamos, NM   87545

Fabio Scarpa
I.N.F.N.-Lab. Naz. Legnaro
Via Romea 4
I-35020 Legnaro
ITALY
49-641200

Brian Schmidt
Physics Department
Florida State University
Tallahassee, FL   32306
(904) 644-6584

James E. Sealey
Sealey Instrument Co. Inc.
P.O. Box 48119
Atlanta, GA   30362

David M. Shepherd
Megavolt Ltd.
Corn Hill
Ilminster
ENGLAND   TA19 0AH
460-57458

William R. Shields
Janis Research Co. Inc.
2 Jewel Drive
Wilmington, MA   01887
(617) 657-8750

William Smith
GMW Associates/Danfysik
P.O. Box 2578
Redwood City, CA   94064
(415) 368-4884

John Southon
Simon Fraser University
GSB105
Physics Dept.
McMaster University
Hamilton, Ontario
CANADA   L8S 4KI
(416) 525-9140 ext. 4046

J.W. Stark
Tandem Laboratory
McMaster University GSB
Hamilton, Ontario
CANADA   L85 4K1
(416) 525-9140 ext. 4046

Michael Stier
National Electrostatic
Graber Rd. Box 310
Middleton, WI   53562
(608) 831-7600

Derek W. Storm
Nuclear Physics Lab/GL 10
University of Washington
Seattle, WA   98195
(206) 543-4085

J. David Sturbois
Ohio University Accelerator Lab
Ohio University
Athens, OH   45701
(614) 593-1983

Suehiro Takeuchi
JAERI
Tokai, Naka, Ibaraki
JAPAN   319-11
081 0292 82 5860

Joseph R. Tesmer
Los Alamos National Lab
P.O. Box 1663
Group P-10, MS-K764
Los Alamos, NM   87545
(505) 667-6370

Tracy N. Tipping
J.R. Macdonald Lab
Physics Dept.
Kansas State University
Manhattan, KS   66506
(913) 532-6782

Roger R. Tremblay
Chalk River Nuclear Lab
Atomic Energy of Canada Limited
Chalk River, Ontario
CANADA   K0J 1J0

Claudio Tuniz
Physics Dept.
Rutgers University
Piscataway, NJ   08855
(201) 932-2404

Jesse L. Weil
Dept. of Physics and Astronomy
University of Kentucky
Lexingotn, KY   40506

Mike Wells
Physics Department
Kansas State University
Manhattan, KS   66506
(913) 532-6782

Chris Westerfeldt
Triangle Universities Nuclear Lab
Duke Physics Dept.
Duke Station
Durham, NC   27706
(919) 684-8271 (8158)

Douglas Will
Physics Lab/GL-10
University of Washington
Seattle, WA   98195
(206) 543-4080

Norval Ziegler
Oak Ridge National Lab
P.O. Box X
Oak Ridge, TN   37831
(615) 574-4761

Gary P. Zinkann
Argonne National Lab
Physics Div. Bldg. 203
9700 S. Cass Ave.
Argonne, IL   60439
(312) 972-4115

Heimart von Zweck
PHYSICON Corp.
221 Mount Auburn St.
Boston, MA   02138